Springer Proceedings in Physics

Volume 235

Indexed by Scopus

The series Springer Proceedings in Physics, founded in 1984, is devoted to timely reports of state-of-the-art developments in physics and related sciences. Typically based on material presented at conferences, workshops and similar scientific meetings, volumes published in this series will constitute a comprehensive up-to-date source of reference on a field or subfield of relevance in contemporary physics. Proposals must include the following:

- name, place and date of the scientific meeting
- a link to the committees (local organization, international advisors etc.)
- scientific description of the meeting
- list of invited/plenary speakers
- an estimate of the planned proceedings book parameters (number of pages/ articles, requested number of bulk copies, submission deadline).

More information about this series at http://www.springer.com/series/361

Marek Kuś · Bartłomiej Skowron
Editors

Category Theory in Physics, Mathematics, and Philosophy

INTERNATIONAL CENTER
FOR FORMAL ONTOLOGY

Springer

Editors
Marek Kuś
Center for Theoretical Physics
Polish Academy of Sciences
Warsaw, Poland

Faculty of Administration and Social Sciences, International Center for Formal Ontology
Warsaw University of Technology
Warsaw, Poland

Bartłomiej Skowron
Faculty of Administration and Social Sciences, International Center for Formal Ontology
Warsaw University of Technology
Warsaw, Poland

ISSN 0930-8989 ISSN 1867-4941 (electronic)
Springer Proceedings in Physics
ISBN 978-3-030-30895-7 ISBN 978-3-030-30896-4 (eBook)
https://doi.org/10.1007/978-3-030-30896-4

This Springer imprint is published by the registered company Springer Nature Switzerland AG
The registered company address is: Gewerbestrasse 11, 6330 Cham, Switzerland

Introduction

Category theory is becoming increasingly popular in science and philosophy. It is not only the language of mathematics, as it is still often believed. It provides new abstract structures, analogies, unifying tools, and a dynamic way of thinking, which differs from widespread set-theoretical thinking. In 2017, an interdisciplinary conference entitled *Category Theory in Physics, Mathematics, and Philosophy* was held in Warsaw. The conference was organized by the International Center for Formal Ontology (Faculty of Administration and Social Sciences, Warsaw University of Technology), the Copernicus Center for Interdisciplinary Studies, the Center for Theoretical Physics (Polish Academy of Sciences), and the Institute of Mathematics (Academy of Sciences of the Czech Republic). During the conference, we were looking for common categorical threads in these three areas of knowledge. The volume we present to the reader is the result of this conference. Category theory is very abstract, and this is why it naturally corresponds to the ways of thinking that are present in fundamentally different fields of science and humanities. The main thought behind this volume is the conviction that category theory has not only unifying power within mathematics itself. It has built bridges between modern physics, mathematics, computer science, philosophy and others. We hope that this volume will convince the reader—even if only as a mere mental experiment—that it is worthwhile in his or her scientific field to try to think without objects, but only with transformations, assuming for a while that to exist in his or her field of research means to remain in a complex network of relationships.

In the opening contribution entitled "Why Categories?" Marek Kuś, Krzysztof Wójtowicz and Bartłomiej Skowron answer the question of why category theory is becoming more and more popular in various fields of knowledge. In particular, we investigate what makes this theory useful in philosophy and physics. It turns out that the ontology offered by categories is responsible for this.

Zbigniew Król in his paper "Category Theory and Philosophy" analyzes the relationship between category theory and philosophy. He addresses many traditional philosophical questions, including the way of the existence of the mathematical object, Platonism in the philosophy of mathematics, the ontology of sets and ontology of categories, and the question of epistemic access to mathematical reality. He also points out that category theory can be used for formal ontology. His philosophical thesis is clear and surprising: category theory is the most "platonic" mathematical theory. With the approval of one of the reviewers, we enclosed an anonymous commentary to this article, which discusses the theses presented by Król.

Krzysztof Wójtowicz in his philosophical paper "Are There Category-Theoretical Explanations of Physical Phenomena?" considers various types of mathematical explanations of physical phenomena and claims that category theory has no explanatory power in the standard sense of the word "explanation." However, as category theory is a truly abstract field, its explanations belong to metaphysics rather than physics. Category theory being rather an abstract "theory of theories" contributes to the understanding of physics on a meta-level, but does not explain in a standard way concrete physical phenomena, as Wójtowicz argues.

In his philosophical paper "The Application of Category Theory to Epistemic and Poietic Processes," Józef Lubacz considers two types of human activity: acquiring knowledge and the sort of activity that results in the creation of artifacts. Lubacz names the first as epistemic activity and the second as poietic activity. Then, he presents his own metaphysics of the process in which he interprets these two types of activities. Since category theory is underpinned not by the metaphysics of substances, but rather by the metaphysics of the process, Lubacz tries to show the potential for applying the notional framework of category theory to epistemic and poietic processes.

Zbigniew Semadeni in his contribution "Asymmetry of Cantorian Mathematics from a Categorial Standpoint: Is It Related to the Direction of Time?" takes up the following problem: It is known that in category theory every phenomenon has its own dual, i.e., every theorem, definition and proof has its own dual version. In order to obtain the dual category, it is sufficient to reverse the direction of the arrows. However, this type of symmetry does not exist in Cantor's mathematics, i.e., in mathematics where objects are sets with certain structures. Semadeni argues that the basis of this asymmetry is the asymmetry of the many-to-one relationship in the notion of a function. The author also points out that the many-to-one thinking could be related to the arrow of time.

Neil Dewar, Samuel Fletcher, and Laurenz Hudetz in their paper "Extending List's Levels" consider the formal aspects of the relation of supervenience, one of the most widely discussed relations in contemporary metaphysics. The authors evoke the formal scheme proposed by Christian List and show how it can be generalized in order to make greater use of category-theoretic tools and ideas. They also show how such an extended approach is useful both in the philosophy of natural sciences and social sciences.

The next three papers contribute to various fundamental problems of physics concerning the structure of the spacetime. The first one, by Krzysztof Bielas and Jerzy Król entitled "From Quantum-Mechanical Lattice of Projections to Smooth

Structure of $\mathbb{R}^4$" attacks the problem of the emergence of exotic smooth structures in four dimensions in category theory terms. The starting point is a categorical reconstruction of the logical structure of quantum and classical propositional structures. In the classic case, such a structure corresponds to Boolean algebra, whereas for the quantum case, the corresponding structure is that of an orthomodular lattice. A natural question concerning the determination of an orthomodular lattice by its Boolean subalgebras is answered in terms of the categorical setting using the concept of the colimit. Bielas and Król show how to transfer such construction into the description of spacetime as a differentiable manifold. As a result, an exotic smooth structure and, consequently, four-dimensionality appear in a natural way.

The second contribution concerning a categorical description of "pathological" properties of the differential structure of spacetime by Michael Heller and Jerzy Król is devoted to singularities preventing prolongation of some timelike curves. The authors suggest employing Synthetic Differential Geometry (SDG), based on category theory and an ensuing intuitionistic logic, to shed new light on the problem. Such an approach adds to the ordinary differential structure of the real line new kinds of differentials and allows the authors to construct a model in which it is possible to "look beyond the singular boundary."

Synthetic Differential Geometry is also a tool employed by Jerzy Król in his contribution concerning Perturbative Quantum Gravity. One way of looking at the problem of quantization of gravity is the following. The right-hand side of the Einstein equations (the energy–momentum tensor) is determined by quantum fields in the spacetime, whereas the left-hand side (the gravity) involves purely classical quantities (the Ricci and metric tensors). An attempt to quantize fluctuations around the flat metric, as it is done with other fields, leads, however, to an incurably nonrenormalizable theory. Król shows how this incompatibility of Einstein equations with quantum matter sources can be formulated and partially solved in spaces described by SDG.

The paper by Nop, Romanowska, and Smith "Category Theory as a Foundation for the Concept Analysis of Complex Systems and Time Series" deals with an abstract description of commonly encountered situation in which we have collections of objects and properties together with a relation that attributing the latter to the former. The triple consisting of objects, properties, and an attribution relation is termed a context. Identification of the set of all properties common to a subset of items and, dually, the set of all items possessing all given properties allows for the introduction of the notion of a concept connected to a given context and a concept lattice as a set of all concepts of a context. The whole construction has a natural category theory formulation. The authors show how this static description may be extended to cases of time-evolving or hierarchical (multilevel) systems and applied, e.g., to the analysis of time series.

It only remains for us to thank all of the reviewers and commentators. We would like to express our sincere thanks to Torsten Asselmeyer-Maluga, Tomasz Brengos,

Samuel Fletcher, Jerzy Król, Christian List, Marcin Łazarz, Jiří Rosický, Zbigniew Semadeni, and Piotr Sułkowski for their helpful comments and reviews of the articles that make up this volume.

Marek Kuś
e-mail: marek.kus@cft.edu.pl

Bartłomiej Skowron
e-mail: b.skowron@ans.pw.edu.pl

Contents

1 **Why Categories?** . . . 1
Marek Kuś, Bartłomiej Skowron and Krzysztof Wójtowicz

2 **Category Theory and Philosophy** . . . 21
Zbigniew Król

3 **Are There Category-Theoretical Explanations of Physical Phenomena?** . . . 33
Krzysztof Wójtowicz

4 **The Application of Category Theory to Epistemic and Poietic Processes** . . . 45
Józef Lubacz

5 **Asymmetry of Cantorian Mathematics from a Categorial Standpoint: Is It Related to the Direction of Time?** . . . 55
Zbigniew Semadeni

6 **Extending List's Levels** . . . 63
Neil Dewar, Samuel C. Fletcher and Laurenz Hudetz

7 **From Quantum-Mechanical Lattice of Projections to Smooth Structure of $\mathbb{R}^4$** . . . 83
Krzysztof Bielas and Jerzy Król

8 **Beyond the Space-Time Boundary** . . . 95
Michael Heller and Jerzy Król

9 **Aspects of Perturbative Quantum Gravity on Synthetic Spacetimes** . . . 105
Jerzy Król

10 **Category Theory as a Foundation for the Concept Analysis of Complex Systems and Time Series** . . . 119
G. N. Nop, A. B. Romanowska and J. D. H. Smith

Contributors

Krzysztof Bielas University of Information Technology and Management, Rzeszow, Poland;
Institute of Physics, University of Silesia, Chorzow, Poland

Neil Dewar LMU Munich, Munich, Germany

Samuel C. Fletcher University of Minnesota, Twin Cities, Minneapolis, MN, USA

Michael Heller Copernicus Center for Interdisciplinary Studies, Kraków, Poland

Laurenz Hudetz London School of Economics, London, UK

Jerzy Król University of Information Technology and Management, Rzeszow, Poland

Zbigniew Król Faculty of Administration and Social Sciences, International Center for Formal Ontology, Warsaw University of Technology, Warsaw, Poland

Marek Kuś Faculty of Administration and Social Sciences, International Center for Formal Ontology, Warsaw University of Technology, Warsaw, Poland;
Center for Theoretical Physics, Polish Academy of Sciences, Warsaw, Poland

Józef Lubacz International Center for Formal Ontology, Warsaw University of Technology, Warsaw, Poland

G. N. Nop Iowa State University, Ames, IA, USA

A. B. Romanowska Warsaw University of Technology, Warsaw, Poland

Zbigniew Semadeni Institute of Mathematics, University of Warsaw, Warsaw, Poland

Bartłomiej Skowron Faculty of Administration and Social Sciences, International Center for Formal Ontology, Warsaw University of Technology, Warsaw, Poland

J. D. H. Smith Iowa State University, Ames, IA, USA

Krzysztof Wójtowicz Department of Logic, Institute of Philosophy, University of Warsaw, Warsaw, Poland

Chapter 1
Why Categories?

Marek Kuś, Bartłomiej Skowron and Krzysztof Wójtowicz

The aim of philosophy, abstractly formulated, is to understand how things in the broadest possible sense of the term hang together in the broadest possible sense of the term. [35]

Mathematical tools are much richer than our everyday intuitions and purely verbal distinctions; they are able to reveal unexpected aspects of reality. [13]

We did not then regard it as a field for further research efforts, but just as a language and an orientation—a limitation which we followed for a dozen years or so, till the advent of adjoint funtors. [26]

Abstract In this article we answer the question of why categories are becoming more and more popular in physics, mathematics and philosophy. The article presents a review of the role of categories in the philosophy of mathematics, in the foundations of mathematics, in metaphysics and in quantum mechanics. Our claim is that category theory is a formal ontology that captures the relational aspects of the given domain in question.

M. Kuś · B. Skowron
Faculty of Administration and Social Sciences, International Center for Formal Ontology, Warsaw University of Technology, pl. Politechniki 1, 00-661 Warsaw, Poland
e-mail: b.skowron@ans.pw.edu.pl

M. Kuś (✉)
Center for Theoretical Physics, Polish Academy of Sciences, Warsaw, Poland
e-mail: marek.kus@cft.edu.pl

K. Wójtowicz
Institute of Philosophy, University of Warsaw, Krakowskie Przedmieście 3, 00-927 Warsaw, Poland
e-mail: kwojtowi@uw.edu.pl

M. Kuś and B. Skowron (eds.), *Category Theory in Physics, Mathematics, and Philosophy*, Springer Proceedings in Physics 235,
https://doi.org/10.1007/978-3-030-30896-4_1

1.1 Introduction

Eilenberg and Mac Lane—the founders of category theory (CT)—initially treated their invention as a useful language for certain mathematical problems. Its intensive and unexpected development in the twentieth century led to the fact that it was an increasingly popular theory among mathematicians, computer scientists, physicists, engineers and philosophers alike. What has made CT so popular and why is it being used more and more widely in so many different fields of knowledge? This review article is supposed to answer—at least partially—this question.

In the second section (following the introduction) we describe the beginnings of CT. The third section deals with relations between CT and philosophy, and in particular with philosophical problems concerning the foundations and unity of mathematics, as well as the problem of structuralism in the philosophy of mathematics. The third section also describes how CT affects contemporary metaphysics. The fourth section deals with the role of CT in contemporary physics, in particular quantum mechanics.

1.2 The Beginnings of Category Theory

Category theory is a joint work of Samuel Eilenberg and Saunders Mac Lane. Their collaboration in the 1940s led to its creation. More specifically, the emergence of category theory was determined by the combination of Mac Lane's algebraic talent and Eilenberg's topological talent. Here, Mac Lane himself recalls the origins of CT [2, p. 20–1]:

> In the spring of 1941 Michigan invited me to give a series of five or six lectures, so I talked about group extensions. This was a subject on which I had done some work and it came out of my earlier work on valuations with [O. F. G.] Schilling. I had calculated a particular group extension for p-adic solenoids. Eilenberg was in the audience, except at the last lecture, and made me give the last lecture to him ahead of time. Then he said, "Well, now that calculation smells like something we do in topology, in a paper of [Norman] Steenrod." So we stayed up all night trying to figure out what the connection was and we discovered one. We wrote our first joint paper on group extensions in homology, which exploited precisely that connection. It so happened that this was a time when more sophisticated algebraic techniques were coming into algebraic topology. Sammy knew much more than I did about the topological background, but I knew about the algebraic techniques and had practice in elaborate algebraic calculations. So our talents fitted together. That's how our collaboration got started. And so it went on for fifteen major papers.

At first, it did not seem that CT would constitute a separate and independent subject of mathematical research. The notion of category was only an auxiliary notion, which was needed for other purposes—it was just an abstract basis for research on the phenomenon of natural equivalences. In their joint article *General Theory of Natural Equivalences* in 1945, which now serves as a classic reference, Eilenberg and Mac Lane [10, p. 247] claimed:

> It should be observed first that the whole concept of a category is essentially an auxiliary one; our basic concepts are essentially those of a *functor* and of a natural transformation (...). The idea of a category is required only by the precept that every function should have a definite class as domain and a definite class as range, for the categories are provided as the domains and ranges of functors. Thus one could drop the category concept altogether and adopt an even more intuitive standpoint, in which a functor such as "Hom" is not defined over the category of "all" groups, but for each particular pair of groups which may be given. The standpoint would suffice for the applications, inasmuch as none of our developments will involve elaborate constructions on the categories themselves.

Eilenberg and Mac Lane introduced very abstract tools into mathematics, which seemed even too abstract. Nevertheless, they motivated their work with both technical merits, which allow for an effective study of the phenomenon of naturality, and conceptual advantages. They noted that the proposed conception is so general that it allows for the detection of the same structures in fundamentally different fields of mathematics. By finding new analogies between different fields of mathematics it suggests new results. Thanks to the fact that categorical glasses allow for the observation of the same structures in both topology and algebra, these glasses allow for a unifying view of mathematics. Already in 1945 it was clear that CT had the power to unify mathematics.

From an ontological point of view, it can be said that Eilenberg and Mac Lane have made a certain shift. Well, mathematical objects in practice are considered as if they were autonomous, separate from other objects. As if they existed as independent substances, whose interior determines what they really are. It is enough to look inside to know what properties they have. This is a standard and natural cognitive approach to mathematical objects. Eilenberg and Mac Lane did it differently, contrary to this natural and widespread attitude. They suggested that mathematical objects should always be considered with their surroundings. If we consider groups, we should consider them together with all homomorphisms, if we consider topological space, we should consider all homeomorphisms. Therefore, we do not consider objects in themselves, but consider them simultaneously with morphisms; in other words, we do not consider individual objects, but *categories* [10, p. 236].

The ontological shift proposed by the fathers of CT has many consequences. Group theory in the categorical approach becomes a study of the invariants of the respective functors. Group theory explores constructions that are covariant or contravariant under homomorphisms. In their words: "group theory studies functors defined on well specified categories of groups, with values in another such category" [10, p. 237]. It was not a completely new approach. The authors themselves have noticed that this is actually an extension of Klein's Erlanger Programme. Geometric space was considered by Klein together with its transformation group, while Eilenberg and Mac Lane suggested that one should consider the categories together with its algebra of morphisms.

The proposal of such a general conceptual framework was not obvious at that time. Mac Lane himself wondered whether categorical concepts had been introduced too early. Here, again, in his own words [26, p. 334–5]:

> It was perhaps a rash step to introduce so quickly such a sweeping generality—an evident piece of what was soon to be called "general abstract nonsense." One of our good friends (an admirer of Eilenberg) read the paper and told us privately that he thought the paper was without any content. Eilenberg took care to see to it that the editor of the *Transactions* sent the manuscript to a young referee (perhaps one who might be gently bullied). The paper was accepted by *Transactions*. I have sometimes wondered what could have happened had the same paper been submitted by a couple of wholly unknown authors. At any rate, we did think it was good, and that it provided a handy language to be used by topologists and others, and that it offered a conceptual view of parts of mathematics, in some way analogous to Felix Klein's "Erlanger programme." We did not then regard it as a field for further research efforts, but just as a language and an orientation—a limitation which we followed for a dozen years or so, till the advent of adjoint funtors.

Nevertheless, such abstract concepts as category hung in the air somewhere at the time. What if Eilenberg and Mac Lane hadn't introduced CT? Mac Lane speculated that other mathematicians would have done it, unless they were afraid of the excessive abstraction of the emerging concepts (cf. [28, p. 210]). Among the potential creators of category theory he listed Claude Chevalley, Heinz Hopf, Norman Steenrod, Henri Cartan, Charles Ehresmann, and John von Neumann (cf. [18, p. 3]).

Category theory developed very quickly and intensively. As one of the breakthrough years for the development of CT Mac Lane [26, p. 346] indicates the year 1963. It was then that Lawvere's groundbreaking dissertation appeared, which contained categorical descriptions of algebraic theories and many other important ideas. That year also marked the first public presentation of the adjoint functor theorem by Freyd, Ehresmann published his paper on what we call internal categories, Mac Lane's first coherence theorem also appeared in 1963. And SGA IV was also published—the seminar notes from the *Séminaire de Géométrie Algébrique du Bois Marie* run by Alexander Grothendieck. As Mac Lane [26, p. 347] estimates between 1962 and 1967 around 60 people started working in category theory. Already in 1965 in California, at the CT conference, Lawvere delivered the talk "The category of categories as a foundation of mathematics" [26, p. 351].

The classic (still advanced) textbook in category theory *Categories for the Working Mathematician* was published by Mac Lane for the first time in 1976. In 1992 Mac Lane, jointly with Ieke Moerdijk, published *Sheaves in Geometry and Logic: A First Introduction to Topos Theory*. In 2017, a textbook for philosophers was published: *Categories for the Working Philosopher*, edited by Elaine Landry [19]. In 2006 Steve Awodey, Mac Lane's last Ph.D. student, wrote a textbook *Category Theory* that was easier to read then the textbook by Mac Lane. That's why Awodey felt the need to write a new textbook [3, p. v]:

> Why write a new textbook on Category Theory, when we already have Mac Lane's *Categories for the Working Mathematician*? Simply put, because Mac Lane's book is for the working (and aspiring) mathematician. What is needed now, after 30 years of spreading into various other disciplines and places in the curriculum, is a book for everyone else.

Let's see what CT looks like in Poland. Due to the intensive development of category theory, students of the first years of mathematical studies can now study CT in some departments in Poland. Categorical terms are often introduced in other courses, e.g. in topological courses. The first (and so far the only) CT textbook in Polish was *Wstęp do teorii kategorii i funktorów* published by Zbigniew Semadeni and Antoni Wiweger in 1972.

1.3 Category Theory and Philosophy

Category theory, as well as set theory and, unlike algebraic topology, aroused the interest of philosophers from the very beginning. Already in the classic paper [10, p. 247] there are comments on the foundations of category theory (but not yet the foundations of mathematics!), in particular there are references to an unramified theory of types or to the Fraenkel–von Neumann–Bernays' system—these systems are mentioned as possible solutions for the ontological foundation of categories.

Multidimensional relations occur between category theory and philosophy.[1] Below we will restrict ourselves to a brief discussion of four selected themes:

1. the discussion between mathematical structuralism and object realism;
2. the problem of foundations of mathematics;
3. the problem of the unity of mathematics;
4. the role of CT in contemporary metaphysics.

1.3.1 Structuralism Versus Object Realism

One of the fundamental philosophical question concerns the nature of mathematical concepts—which, under the realistic interpretation, is formulated as the ontological question concerning the status and properties of mathematical objects. Speaking in very general terms, the question is, whether mathematical objects have any intrinsic properties, or whether their properties are purely relative. Earlier we pointed to this issue when we mentioned the ontological shift of Eilenberg and Mac Lane. A related question concerns the identity criterion for mathematical objects: is identity determined by some immanent properties of the objects, or rather purely by the relations in which this object stands to other mathematical objects?

Mathematical structuralism rejects the view, that there are intrinsic properties. According to the structuralist point of view, an object is constituted by the relations to other objects. For instance, the function of a president (of a country or a company) is defined regardless of the particular person holding the office—and only due to the place within the whole structure (country or company).

Consider natural numbers. According to the structuralist picture, natural numbers can be characterized only by their role in the natural numbers structure (i.e. in the ω-sequence). They have no intrinsic properties, and what really matters are only

[1]In this volume, see [8, 17, 23, 37, 44].

relationships between natural numbers. Indeed, the question, whether the number 5 has its properties regardless of the existence of other natural numbers seems strange—for instance, whether it is still prime. The notion of primeness would not make sense if only one number existed. This way of thinking is then extended to other mathematical objects: according to this view, the only properties of mathematical objects are relational properties—so their identity is determined only via the roles they play in mathematical structures. We can say, that mathematical objects are like vertices in a giant graph or—as Mac Lane (see [24, Chap. XII entitled *The Mathematical Network*]) would probably say—network.

The argument between structuralist and non-structuralists is a prototypical example of an relativist-essentialist argument in metaphysics [34, p. 86]:

> Quine has captured the incompleteness of mathematical objects in his doctrine of ontological relativity: there is no fact of the matter as to whether the ontology of one theory is included in that of another except relative to an interpretation of the former in the latter. What I have tried to do so far is show that Quine's surprising doctrine is what we would expect to hold in mathematics.

The categorical point of view seems to be particularly well-suited to express the intuitions of mathematical structuralism. Take a typical expression of the structuralist position [34, p. 84]:

> For me, mathematical objects have no distinguishing characteristics except those they have by virtue of their relationships to other positions in the structures to which they belong. In short, I take the geometrical point, the paradigm position, as a paradigmatic mathematical object.

The familiar illustration of categories in terms of graphs (with black dots and arrows) suits this view very well. All the black dots look alike—and the only things, that determines their identity is the place within the graph, in particular the arrows which connect the dot with the other dots. The properties of objects within a category are defined only via the respective morphisms. For instance, the "essence" of being the initial or terminal object is captured by the morphisms—not by any immanent properties, which are secondary. Also the notion of identity in some sense loses its meaning, as we only speak of isomorphisms within a category (and this is obviously a category-relative notion).

Objects and morphisms within the category live on the first level of abstraction. They might be considered to be natural environment(s) for different kinds of mathematical objects. But what is much more interesting and conceptually fertile are the relationships between categories, expressed in terms of functors (and higher-order constructions, like natural transformations). Categories—in a sense—become "dots" and are viewed from a higher level.

Coming back to the natural numbers example we can not only think of the particular numbers as points ("dots") within a structure (so that the "dots" inherit their properties from the structure they live in). We can also think of the natural number structure in terms of its universal properties within a "higher-order structure", i.e. the Peano category, as Mazur [29, p. 231–2] claims:

> This strategy of defining the Natural Numbers as "an" initial object in a category of (what amounts to) discrete dynamical systems, as we have just done, is revealing, I think; it isolates, as Peano himself had done, the fundamental role of mere succession in the formulation of the natural numbers.

A paradigm example of a structure with purely relational properties is a group. It does not make sense to ask, what the neutral element "really is": of course, a group can have many representations (as number, matrices, linear transformations—and many others), but the "essence" of the neutral element is exactly being neutral with respect to the operation (and not being a 15×15 identity matrix or a certain function).

1.3.2 The Foundations of Mathematics

The problem of finding a suitable foundation for the mathematical edifice has been discussed extensively for at least 150 years—and the crisis in foundations of mathematics around 1900 made the topic very hot.[2] There are many mathematical disciplines, which *prima facie* seem very different—like geometry and algebra in the historical sense of these terms (today they are of course "entangled" in a profound way, and the term "algebraic geometry" illustrates that). Do these different disciplines have common roots?

According to the most widespread view, set theory can serve the role of a foundational theory. Indeed, set theory is so strong and general that virtually all mathematical notions previously used by mathematicians in an informal way (e.g. the concepts of: natural, rational, real, complex numbers; real-valued function; probability; differentiation in the real and complex sense; Banach space; differential manifold, etc.) can be formally reconstructed in the language of set theory. But there is a tension between the fact, that mathematical notions can be formally represented in set theory—but on the other hand, that they have a meaning outside the context of set theory, and lead a happy life without ever noticing, that (according to some foundationalists) they are really sets. In fact, mathematical practice does not really need set theory—apart from some elementary textbook facts.[3] More advanced set-theoretic notions (large cardi-

---

[2]Perhaps today the discussions are not so emotional, and there is no crisis in sight: even if some philosophical doubts can be formulated, mathematics seems to be doing well even without any (official) foundations. But this is more of a pragmatic than a fundamental issue.

[3]The problem of how much set theory is needed, and whether this set theory is really "set-theoretic set theory" or perhaps "category-theoretic set theory" is discussed for instance by Colin McLarty in [30, p. 2]:

> All support the claim that mathematicians know and use the concepts and axioms of the Elementary Theory of the Category of Sets (ETCS), often without knowing or caring that they are the ETCS axioms.

He examines the example of two standard and influential textbooks on topology and algebra. McLarty also claims:

nal, Boolean-valued model) are very remote from everyday mathematical practice and play virtually no role there.

A standard illustration are natural numbers. Usually, on the pre-theoretic level, we consider them to be objects per se—and the computations performed on them have a "self-contained character". But from the point of view of the set-theoretic foundations of mathematics, natural numbers are just sets. For instance, in the (rather standard) von Neumann representation, natural numbers are identified with finite ordinal numbers.[4] But there are many representations of this kind, and all of them are admissible, from the logical point of view.[5] But from the point of view of the working mathematician (for instance, a number theorist), the set-theoretic reduction of numbers plays no role, as mathematicians obviously do not think of natural numbers as being sets obtained from an empty set of set-theoretic operations! Indeed, the question, whether $0 \in 2$ seems rather awkward, and is surely irrelevant for, say, the Twin Primes Conjecture. So the problem of set-theoretic reduction seems artificial from the point of view of mathematical practice: no mathematician is really worried about the problem of translating the results into some awkward logical notation (into the language with one two-place predicate $\in$, i.e. into set theory).

The foundational enterprise has a theoretic character, just like the formalizability postulate: no mathematician bothers with formalizing proofs to the full extent, however, this postulate (the common belief, that proofs can—in principle—be formalized) serves as a kind of methodological warrant. Complaints about the artificial character of the reductions can be countered with the observations, that our everyday habits are not a warrant for methodological correctness. We are free to use informal, and even metaphorical language even in science—as long as we are aware of the fact, and are able to provide rigorous paraphrases.

Category theory is believed (perhaps not very widely—but very firmly!) to be a good candidate for a foundational theory. But the notion of "foundation" is different than in the case of set theory. It is not about isolating one single notion or a theory (like "set" and "set theory") which appears in the ultimate definiendum of all mathematical notions. This point of view is typical for set-theoretic foundationalism. Adherents of category-theoretic foundations stress the fact, that CT takes mathematics "at face value": mathematical notions are not forced into the Procrustean bed of set theory (or some other formal theory we choose), but are free to live within their natural conceptual environment: groups populate the category Grp; topological spaces live within Top; partial orders also have their own place to live. Sets are

---

the category of sets described in ETCS is a closer fit to the practical needs of most mathematicians than is the cumulative hierarchy of sets described in ZFC.

[4]The empty set becomes 0; the singleton $\{\emptyset\}$ becomes 1; $\{\emptyset, \{\emptyset\}\}$ plays the role of 2, *etc*. The successor functions is defined as a set-theoretic operation: $n + 1 = n \cup \{n\}$, and the resulting sequence of natural numbers is: $\emptyset, \{\emptyset\}, \{\emptyset, \{\emptyset\}\}, \{\emptyset, \{\emptyset\}, \{\emptyset, \{\emptyset\}\}\}, \ldots$

[5]An example of a different representation is: $\emptyset, \{\emptyset\}, \{\{\emptyset\}\}, \{\{\{\emptyset\}\}\}, \{\{\{\{\emptyset\}\}\}\}, \ldots$—i.e. the empty set is 0, and the rule for successor is: $n + 1 = \{n\}$. This leads to the famous multiple-reduction problem in philosophy of mathematics (the *locus classicus* is [5]). As there are many possible reductions, it is not clear, that any of them is the proper one. So perhaps identifying numbers with sets is not legitimate.

indeed important—they have their category of their own—but this is just one of many categories, by far neither the only, nor the most important one. The notion of set is one of many natural mathematical notions—just like the notion of continuity. And even the logic is not absolute—a topos might have some internal logic, and could serve as a way of interpreting mathematical notions within it.[6]

So this way of thinking is not quite reductionist—it might rather be considered as providing a useful, enlightening template for interpreting mathematical notions—within their respective domains. This constitutes a strong shift in thinking about mathematical notions: what is most interesting about them, are their universal properties. A tensor product is important primarily because it is an object representing an important functor, not interesting per se. Even the Cartesian product of two sets is rather viewed as an object with appropriate arrows—and this is important, not the particular ordered pairs.

An important general insight is therefore perhaps: mathematics does not need foundation—but organization. As Mac Lane [24, p. 406] put it:

> Alternatively, set theory and category theory may be viewed as proposals for the organization of Mathematics. The canons of set theory provide guides to the formulation of new concepts and emphasize the extensional character of Mathematics: A “property” is completely determined by knowing all the elements which have that property. Similarly, the canons of category theory emphasize the importance of considering not just the objects but also their morphisms. They also emphasize the use of universal constructions and their associated adjoint functors.

For Mac Lane, none of these proposals are fully successful. Category theory works well in algebra and topology, but not so well in analysis. Set theory, for Mac Lane, contains many artificial constructions and, as Mac Lane [24, p. 407] has repeatedly said following Hermann Weyl: “it contains far too much sand”.

CT serves as a method for organizing mathematics rather as a foundation. Nevertheless, CT still has ambitions to replace set theory’s pride of place. In recent years, homotopy type theory and univalent foundations (HoTT/UF) have been intensively developed and has become a serious competitor to set theory. In principle, HoTT/UF can serve as a foundation for mathematics. It is a paradigm that marks a new way of thinking about the foundations of mathematics, which more faithfully than set theory, represents everyday mathematical practice (in this respect CT has always challenged set theory) on the one hand, and on the other HoTT/UF “is suited to computer systems and has been implemented in existing proof assistants” [42, p. 7].

In homotopy type theory the idea of a collection is realized by a type, just as in set theory the idea of a collection is realized by a set. The elements of types are points. A set is made up only of elements, but the type is made up of both points and ways of identifying points. Two points can be the same in many ways (let’s take the sets $\{a, b\}$ and $\{c, d\}$ as the points and bijection as a relation of being the same, in this case there is more than one bijection; or: two topological spaces can be homeomorphic in many ways), so types also consist of ways in which elements are the same.

---

[6]This problem is discussed in the context of Topos Quantum Theory in [44].

Types are therefore certain spaces, not pure sets, more specifically they are ω-grupoids. In set theory the relation of being the same is given a priori, from the very beginning, as being in a sense ready. In homotopy type theory, if a collection is considered, reasons for "being the same element" must be considered simultaneously. For the foundations of mathematics it is important that the universe of sets is an appropriate part of the universe of types (through ETCS), therefore the model of ZFC can be reconstructed in HoTT/UF. Thus, if set theory reconstructs mathematical objects, then HoTT/UF does it all the more (for a more detailed explanation see a brief discussion of the idea in [38] and a full presentation of HoTT/UF in [42]). Undoubtedly HoTT/UF is much richer—ontologically speaking—than set theory [42, p. 1]:

> Homotopy type theory also brings new ideas into the very foundation of mathematics. On the one hand, there is Voevodsky's subtle and beautiful *univalence axiom*. The univalence axiom implies, in particular, that isomorphic structures can be identified, a principle that mathematicians have been happily using on workdays, despite its incompatibility with the "official" doctrines of conventional foundations. On the other hand, we have *higher inductive types*, which provide direct, logical descriptions of some of the basic spaces and constructions of homotopy theory: spheres, cylinders, truncations, localizations, etc. Both ideas are impossible to capture directly in classical set-theoretic foundations (...).

1.3.3 The Problem with Unifying Mathematics

> *Category theory takes a bird's eye view of mathematics. From high in the sky, details become invisible, but we can spot patterns that were impossible to detect from ground level. [31]*

Notwithstanding the fact, that there are dozens of mathematical disciplines (and hundreds of subdisciplines), it is an empirical datum, that mathematics is a unity. Notions, methods, theorems are transferred in a very natural way from one mathematical discipline to another; examples are abundant. But in the history of mathematics, these theories often emerged as separate disciplines, and great contributions in the history of mathematics consisted often in transferring concepts from discipline D_1 to discipline D_2—applying algebra in geometry to solve classic questions (like the angle trisection) is the most obvious illustration. This mutual applicability of mathematical disciplines is almost a "raw data", a phenomenon which requires an explanation. The insights from CT might provide it.

From the point of view of reductive foundationalism (for instance—set theoretic), the answer is very natural: mathematics is a unity just because of the fact that all of Mathematics can be reduced to the Fundamental Theory (whatever it is). But this is not exactly what mathematical practice suggests: the mutual applications of diverse mathematical notions and results is not the result of translating everything into set-theoretic language, and then proceeding within set theory. The conceptual links are much more direct, that via a set-theoretic translation.

So, it is natural to ask, whether the phenomenon of unity could be explained not by postulating a reduction to set theory, but in a more direct way. This is the point of view of CT: the unity of mathematics stems from the fact that all mathematical theories and objects live within their natural environment(s) (which are the respective category/ies)—and the unity is explained by investigating the relationships between these categories. From this point of view, some notions become natural, and "environment-relative"—one of the simplest examples is the notion of isomorphism. In set theory there is an enormous amount of the "implementations" of this notion (isomorphism of groups is something different than the isomorphism of rings, even if the underlying set is the same[7]). In CT the isomorphism between objects in a category is defined via the properties of the morphisms. And the "transfer" of isomorphisms between different categories is explained via functorial notions (like the homeomorphism of topological spaces and algebraic isomorphisms). We might say, that the "essence" of the familiar notion of isomorphism (in all its variants) is captured by one category-theoretic definition.

In the development of mathematics, isolating proper, core notions for a discipline (e.g. by finding appropriate primitive notions for an axiomatization) was a natural problem. Identifying such notions can have great explanatory value, in a sense—metaphorically speaking—this is about identifying "the essence of the theory". And the proper identification can lead to a fertile conceptual recasting—being an explanatory presentation of a discipline.

This statement is perhaps difficult to grasp in a precise way: what does it mean, that a theory is presented in an explanatory or a non-explanatory way? An illuminating example is presented in [32], where Pringsheim's presentation of complex analysis is discussed.[8] According to Mancosu [32, p. 108]:

> The original approach to complex analysis defended by Pringsheim is based on the claim that only according to his method it is possible to "explain" a great number of results, which in previous approaches, in particular Cauchy's, remain mysterious and unexplained.

Of course, this is not a new theory, there are no new theorems—rather, this is an example of a shift in perspective. Mancosu's example does not concern CT in any way—but illustrates the phenomenon. Providing a reformulation of a theory can provide important insights. And the problem of "immersing" different subjects into one conceptual system is obviously connected with questions concerning the unity of a subject.

At this point it is worth mentioning that CT in the person of F. W. Lawvere led to a peculiar demythologisation of Gödel's famous theorems. It turns out that both Gödel's incompleteness theorem and Russell's paradox, as well as Cantor's theorem and Tarski's undefinability theorem—all these results are instances of a simple categorical construction. An interested reader should take a look at Lawvere's work in [21], and a reader who is not familiar with CT yet can easily follow Lawvere's result, thanks to N. Yanofsky's accessible introduction to the subject in [45] without the use of a CT-toolbox. This observation by Lawvere is a good example of how CT naturally

[7]Isomorphism of Lie groups *as* groups is something quite different than their isomorphism *as* topological spaces.

[8]Without going into details, it takes the notion of the mean value of a function as basic—and the other (standard) definitions are treated as derived facts.

finds surprising similarities between different mathematical phenomena. It is probably even more surprising that the topological operator of the interior is connected with the inclusion operator in the same way as the existential quantifier with the general quantifier—these are the examples of adjoint functors that were noticed in principle only thanks to CT. An overview of many examples of adjoint functors can be found in Mac Lane's classic work *Categories for the Working Mathematician* [27]. The issue of adjointness in the foundations of mathematics (understood not as "starting-point" or "justification" but as the study of what is universal in mathematics) is discussed by Lawvere in his work *Adjointness in Foundations* [20]. Lawvere shows how one can understand the game of *Formal* and *Conceptual* aspects of mathematics in the context of their adjointness.

CT is not an object-level theory, it is perhaps rather "a theory of theories"—and its contributions to our understanding are at a quite abstract level. As Spivak [41, p. 400] put it:

> Category theory is not a theory of everything. It is more like, as topologist Jack Morava put it, (...) "a theory of theories of anything". In other words, it is a model of models. It leaves each subject alone to solve its own problems, to sharpen and refine its toolset in the ways it sees fit. That is, CT does not micromanage in the affairs of any discipline. However, describing any discipline categorically tends to bring increased conceptual clarity, because conceptual clarity is CT's main concern, its domain of expertise. (...) Finally, category theory allows one to compare different models, thus carrying knowledge from one domain to another, as long as one can construct the appropriate "analogy", i.e., functor.

And this might be characteristic of CT: it takes on a new perspective. This new perspective is indeed an ontological shift that we mentioned earlier. It is a change of ontological form, whereby form, following Ingarden (in his *Controversy over the Existence of the World* [15]) we mean something radically non-qualitative. The form can be a parthood and a wholehood, a substance and its properties (the form here is the subject of properties) or exactly relationality, as in the case of CT. Existence in CT is only and exclusively being in relation. Hence, ontologically speaking, Eilenberg and Mac Lane made a formal-ontological shift towards pure relationality. This is not a technical (in the mathematical sense) shift, many mathematicians use both CT and set theory simultaneously. However, the difference between CT and set theory lies in the fundamentally different ontology behind them, hence discussions between fierce category theorists and set theorists resemble discussions between metaphysicists. CT is essentially a formal metaphysics of mathematics that defends a different vision of the world.

1.3.4 Category Theory in Metaphysics

Spivak [41, p. 382] pointed out that CT has served science as a modeling language for various studied phenomena:

> There is a good deal of work on using category theory to model high-level conceptual aspects of scientific subjects. For example, categories have been used by John Baez to model

> signal flow and reaction networks, by Abramsky and Coecke to model aspects of quantum mechanics, and by Lambek to model computer programming languages. At the time of writing, I believe we are at the early stages of an effort to "categorify" science.

Spivak explaining his proposal—through persuasive examples—claims that CT is a kind of "mathematical model of mathematical models". Let's follow Spivak with the linearity phenomenon. The vector space $\mathbb{R}$ models line-hood, $\mathbb{R}^2$ is a model of plane-hood, etc. Hence, every object in the category of vector spaces $Vect_{\mathbb{R}}$ is a model of linearity. CT, on the other hand, models the relationships between these models. The category of vector spaces $Vect_{\mathbb{R}}$ is also a model, but of a higher order—it is a model, according to Spivak's proposal, of linearity *itself*. Objects of a given category, i.e. low-level models, are in fact determined by the relations between them, these relations are captured by CT by studying the category $Vect_{\mathbb{R}}$. Of course, in the category Cat, composed of all small categories, the phenomena of higher levels are modeled. In particular, in the Cat there are two categories that represent object-hood and morpism-hood in the way that $\mathbb{R}$ models line-hood.

CT in fact is not the only language to model higher level entities, as Spivak initially claimed. In fact, CT in his approach is a higher-level formal ontology, i.e. it is a kind of ontological material that allows us to model many phenomena—and it is much more than just language. Spivak adopts the Kantian perspective and treats the models as cognitive products that do not have an independent existence outside our cognitive apparatus. It is no mystery that this is not the only possible solution. The category theory can be seen as a pre-configuration of the necessary and possible combinations of content, or as a network of ideas (an idea in a sense similar to Plato's idea and the conception of an idea coined by phenomenologists such as Jean Hering and Roman Ingarden [15, p. 67–74]) that determines the possible configurations. Taking this position, Spivak's models become concretizations of pure ideal qualities in the contents of ideas, not just mental representations. However, regardless of ontological differences, the practice of using CT—adequately described by Spivak—is simply common in scientific practice.

The CT thus serves as a modeling tool not only for science but also for philosophy, which is perhaps more important for philosophy itself, because it can lead to its development (if any development at all would be allowed—some philosophers complacently claim that practicing philosophy is like shaving every day, nothing constructive comes from this, except that you are shaved). We will provide two examples here of the use of categorical structures as models in philosophical considerations.

As is well known, Plato has divided reality into two parts, into what really exists (*ideas* or *forms*) and into that which only imitates (*things*). At least since then there has been a dispute in philosophy about whether and how ideas exist and what is their relation to the world of real objects. In the twentieth century, Jean Hering and Roman Ingarden significantly developed their theory of ideas. These philosophers have worked in a phenomenological tradition that is focused on analyzing the content of what appears in direct experience. Of course, this method, on the one hand, allows us to focus on the things themselves (phenomenologists have spoken of a coming back to the things themselves), but on the other hand, it excludes many other operations, such as modeling in Spivak's view or the paraphrase method in Ajdukiewicz's view (cf. [1]), as well as many others. Mathematical analysis, including mathematical modeling, has also entered into philosophy, often called mathematical philosophy, which is promoted e.g. by Hannes Leitgeb (see overview in [22]). Skowron [40], who tries to defend the contemporary theory of ideas against its critics, has taken advantage of this opportunity and for this purpose he uses CT.

Many heavy accusations have been made against the theory of ideas since its inception. Aristotle himself formulated many objections. He claimed that it is problematic to create many ideas from one (One-Over-Many Argument), he asked whether the negations of ideas are also ideas (if there is an idea of man, is there an idea of a non-man?—Negation Argument), he repeated after Plato the Third-Man Argument which supposedly ridicules the participation of ideas in things. He stated that "[T]o say that they are patterns and the other things share in them is to use empty words and poetical metaphors" (Metaphysics, I, 997b 5–12). Using a slightly different modeling method than the one proposed by Spivak, i.e. an improved method of Ajdukiewicz's paraphrases, Skowron rejects these classical arguments against the theory of ideas. He analyzes the phenomenon of negation using CT in joint paper with Wiesław Kubiś [39], by using the notion of n-category he rejects the Third-Man Argument, and he is using elementary categorical constructions to reject the One-Over-Many Argument. CT therefore also serves pure metaphysics by providing conceptual and argumentative tools that were not previously available to philosophers, and thus enriches metaphysical argumentation.

The second example of CT applications in metaphysics is a categorical representation of Leibniz's monadology, proposed by Michał Heller. He presents a toy-model of Leibniz's monadology. Let $Leib$ be a category with an infinite number of objects. We can think of these objects as spacetime events. In this category, each object is connected to each other with only one arrow. Each object in this category is both a terminal and an initial object (so it's a null object). It is known that a terminal object is a unique up to isomorphism, hence all objects in this category are isomorphic. Heller then makes analogies: let's think of objects within the category $Leib$ as Leibniz's monadas, then we obtain the following dependencies [14, p. 194–5]:

- Leibniz claims that every monad is "a perpetual mirror of the universe" or "each substance [monad] 'expresses' the entire universe". In $Leib$ every object is a generalized element of any other object (there is an arrow from every object to every object).
- Leibniz: Monads enjoy a kind of "self-sufficiency", they have "no windows", "with nothing 'entering or leaving', they are 'the sources of their internal actions'": in spite of this "the entire system works harmoniously" ("preestablished harmony"). In $Leib$: all objects are isomorphic (are, in a sense, one object).
- Space is entirely relational, or "merely relative; it is 'order or relation', without related bodies it is 'nothing at all'". In $Leib$ there is nothing except arrows from any object to any other object.

Of course, this is only a model, but the study of its properties may lead to the discovery of further dependencies present in the metaphysics of Leibniz.

1.4 Physics: Category Theory as an *Ontology Rescuing* Tool

Category theory entered physics seemingly only as a new language for old theories. But, in fact, one of the original purposes of the categorical (topos) approach to quantum mechanics, as presented in seminal papers of Christopher Isham, Andreas Döring, and Jeremy Butterfield has been to find an alternative *logical* foundation for the orthodox quantum theory [9, 16]. The most fundamental result of such a shift can be briefly explained by the following observations. The classical and quantum descriptions of the physical world differ considerably on the mathematical level. Classical systems are described in terms of a *phase space*, usually a differential manifold, its (measurable) subsets, coordinate systems, etc. Observables, i.e. physical quantities that we can measure, or, in general to which we can ascribe certain numerical values characterizing the observed system, are functions on the phase space. Observables, such like positions, momenta, energies, angular momenta etc. are some properties of systems like particles, ensembles of particles, rigid bodies, etc. They can change in time, but are properties that are possessed by systems alone and do not depend on whether or not they are actually measured at a particular moment. Moreover, at least in principle, we can measure them without disturbing them. Consequently, measurements can be performed in an arbitrary order, or even simultaneously, and provide the same results. We can thus pose questions about *exact* values of, say, the position and the momentum of a particle. Usually, however, due to e.g. inaccuracies of measurements we inquire into the probability that our particle is in a certain subset of the phase space. Such a probability is determined by the volumes of the relevant subsets.

Quantum mechanics offers a completely different picture. Here we do not have a phase space in the form of a manifold. Instead a system is described in terms of vectors and operators in a Hilbert space. We may ascribe to each system some properties that pretend to be the quantum analogues of classical ones like positions, momenta, angular momenta, energies, etc. (and some others that seem to be of a purely quantum mechanical nature, like spin, isospin, strangeness, hypercharge etc.). However, they are no longer *intrinsic* in the classical sense. They are not "carried" by a system during its evolution, rather they are "brought to life" by an act of measurement, which can be interpreted as an impossibility of a non-disturbing experiment. Each act of measurement disturbs the actual state of a system by bringing it to another state corresponding to a result of the measurement performed. Hence, the order in which measurements are taken does matter, and some measurements can not be taken simultaneously (the uncertainty principle). Moreover, although results of measurements depend on the actual state of the system prior to the act of a measurement, they do it only in a probabilistic manner. This is because for each observable (position, momentum, angular momentum, energy, spin, etc.) there is a corresponding selfadjoint operator, the eigenvalues of which are possible outcomes with probabilities depending on the state of a system before the measurement and the eigenvectors determine possible states after it.

It is thus clear that it is rather hard to find a unifying ontological basis for classical and quantum physics. The ontological status of such fundamental elements of physical reality, as positions, momenta, angular momenta, etc. have radically different ontological status in both theories. Whereas they are intrinsic and objective properties of a physical system, it is not so in quantum theory.

From a purely physical point of view this is not a danger. Ultimately, physics is an experimental science. It can and should answer experimental questions about outcomes of various measurements. Such an approach clearly puts more emphasis on the epistemology, moving apart, or even totally discarding ontological issues.

As an attempt to unify classical and quantum physics on common epistemological ground one can treat the quantum logic approach that goes back to Birkhoff and von Neumann [6]. The main idea is to analyze the structure of elementary experimental question/propositions about a system. In classical physics, elementary propositions can be reduced to statements that values of observed quantities (coordinates) belong to a certain subset of the phase space. The logical structure of the set of such propositions, determined by the rules concerning their negations, conjunctions and disjunctions isomorphically reflects the Boole algebra structure of the set of (measurable) subsets of the phase space. One of the characteristic features of a Boolean structure is the distributivity law, allowing for the distribution of conjunctions over disjunctions and vice versa.

In quantum mechanics elementary propositions concern positions of state vectors (characterizing a state of a system) with respect to eigenspaces of observables (self-adjoint operators in Hilbert space). As in the classical case we can ask composite questions corresponding to conjunctions and disjunctions. However, the ensuing logical structure is no longer distributive. The logic of a system described by a Hilbert space $\mathscr{H}$ is represented by the orthomodular lattice of closed subspaces in $\mathscr{H}$. The involution sending a subspace to its orthogonal complement represents logical negation, satisfying the law of an excluded middle: measuring the spin of an electron will yield either 'up' or 'down', *tertium non datur*. As said, the resulting lattice is non-distributive: x-spin up does not imply x-spin up and z-spin up or x-spin up and z-spin down (the incompatibility of the two measurements is reflected by the non-distributivity of the sub-lattice they 'generate', just as by the non-commutativity of the corresponding sub-algebra of operators). Having the lattice stand for the logic of the system, one derives its probability theory where states assign 'probabilities' to elements of the lattice, respecting the underlying structure (order and complementation). These states turn out to coincide with the usual density matrices by a celebrated theorem of Gleason (as long as $\dim \mathscr{H} \geq 3$, [11]).

Despite differences in the logical structures of both theories, such an approach definitely provides a unifying picture for the whole physics. The differences themselves reflect precisely the dissimilarities between the two theories. Does it really mean that we achieved the goal and we can look at classical quantum physics from the same point of view? For supporters of the purely epistemological approach briefly described above, most probably yes. But for those who pay more attention to the ontological basis of physical theories it might be disappointing. It seems that category theory could be called to come to the rescue.

In a certain sense, one of the original purposes of the categorical (topos) approach to quantum mechanics was to find a well-behaved phase space for a quantum system, restoring, at least to some extent, the well understood ontological basis of classical physics. In addition, it provides at least partial embodiment of Bohr's idea that the account of all evidence concerning quantum phenomena, despite their non-classical character, must be expressed in classical terms [7].

This goal is achieved by replacing the usual non-distributive orthomodular lattices of Birkhoff and von Neumann with the distributive logic of a, so-called, point-free space. Distributivity comes then at the price of the absence of the law of the excluded middle; this is however a feature rather than a flaw, and a characteristic of the true logic of physical observation (see [43]).

The first step along the way is to construct a frame, i.e., a complete distributive lattice, where finite meets distribute over arbitrary joins. The lattice of open subsets of a given topological space provides an example of a frame, moreover one may view any frame as a 'pointless topology', i.e., a virtual space examined only through its collection of open subsets and the lattice operations on them. This does seem to correspond to the way a physicist observes the phase space of a system—and thus to the actual logic of such observations. The use of open subspace reflects the intuitionism of the logic: negation corresponds to taking the interior of the complement, whence the disjunction of a proposition and its negation need not be true. To emphasize this interpretation, one defines the category of locales as the opposite of the category of frames (recall that the topology functor from topological spaces to frames is contravariant).[9]

The whole program can be thus looked upon as a kind of promised restoration of underlying ontology of both classical and quantum theories in the phase space by describing it in terms of topoi (sets in the classical and locales in the quantum case). It can also can be looked upon as a kind of a, previously mentioned, "ontological shift". This time, however, the shift does not enrich the ontology, as it was in the case of mathematical object, but rather impoverishes it, e.g. by denying objective existence, or at least a primary ontological character, to some properties like positions and momenta, both in the quantum and the classical picture. This is the price we pay by an ontological unification of both within the category theory.

The power of such an approach is further exhibited by the fact that it applies also to other theories, different from classical and quantum ones that, to some (limited) extent can be treated as generalizations of quantum mechanics, the so called non-signaling theories [33]. Also in this case a generalized phase-space of the type outlined above can be constructed resulting in a common representation encompassing non-signaling, orthodox quantum, and classical systems and potentially interesting in-betweens [12].

---

[9]Actually, the new intuitionistic logic of the quantum system is realized not as a frame (or any partially ordered set), but rather as a frame object in a suitable topos, intrinsically associated with the system under consideration. Thus, the meta-logic describing the logic of the system is the internal logic of a topos. For instance, probability valuations on the frame object are viewed as morphisms into a real numbers object.

1.5 Conclusion

The emergence and development of category theory was a kind of formal-ontological shift. The standard and natural attitude towards the objects, as if they were individual subjects of properties (as if they were substances and its attributes), was replaced by a form of pure relationality. Existence which was somehow contained in the object was replaced by the necessary coexistence with other objects. CT detached existence from the existing subject, and this is the reason why it still raises some controversy—for some scholars it is still just abstract nonsense.

This new form of relativity is not pure relativism, as one might think at first glance. CT, on the one hand, got rid of the internal structure of objects, replacing it with an appropriate place in the network surrounding the object, and on the other hand, it reached the universal mapping properties, which, being an ontological artifact of CT, is still awaiting an adequate ontological analysis. Didn't CT, which in the person of Saunders Mac Lane had a non-platonic attitude at its core, reach the most platonic objects in modern mathematics?

Acknowledgements The preparation of this paper was supported by an National Science Centre grant, number 2016/21/B/HS1/01955.

References

1. K. Ajdukiewicz, A semantical version of the problem of transcendental idealism (originally published in 1937), in *The Scientific World-Perspective and Other Essays, 1931–1963*, ed. by J. Giedymin (Springer, Netherlands, 1977), pp. 140–154
2. G.L. Alexanderson, A conversation with Saunders Mac Lane, (interview). Coll. Math. J. **20**(1), 2–25 (1989)
3. S. Awodey, *Category Theory* (Clarendon Press, Oxford, 2006)
4. S. Awodey, Structuralism, invariance and univalence, in: [19], 58–68
5. P. Benacerraf, What numbers could not be. Philos. Rev. **74**, 47–73 (1965)
6. G. Birkhoff, J. von Neumann, *The logic of quantum mechanics* (Springer, Berlin, 1975)
7. N. Bohr, in *Albert Einstein: Philosopher-Scientist, The Library of Living Philosophers*, vol. 7, ed. by P. A. Schilpp (Open Court, Evanston, IL, 1949), pp. 200–241
8. N. Dewar, S.C. Fletcher, L. Hudetz, Extending List's Levels, in this volume
9. A. Döring, C.J. Isham, A topos foundation for theories of physics: I. Formal languages for physics. J. Math. Phys. **49**, 053515 (2008)
10. S. Eilenberg, S. Mac Lane, General theory of natural equivalences. Trans. Am. Math. Soc. **58**, 231–294 (1945)
11. A.M. Gleason, Measures on the closed subspaces of a Hilbert space. J. Math. Mech. **6**(6), 885–893 (1957)
12. J. Gutt, M. Kuś, Non-signalling boxes and Bohrification. arXiv:1602.04702
13. M. Heller, Analogy, Identity, Equivalence. Complexity and Analogy in Science, Pontifical Academy of Sciences. Acta **22**, 257–266 (2014). www.pas.va/content/dam/accademia/pdf/acta22/acta22-heller.pdf
14. M. Heller, Category Theory and the Philosophy of Space, in *Filozofia matematyki i informatyki*, ed. by R. Murawski (Copernicus Center Press, Krakow, 2015), pp. 185–200
15. R. Ingarden, *Controversy over the Existence of the World*, vol. I (Peter Lang D, Frankfurt am Main, 2013)

16. C.J. Isham, J. Butterfield, A topos perspective on the Kochen-Specker theorem: I. Quantum states as generalised valuations. Int. J. Theor. Phys. **37**, 2669–2733 (1998)
17. Z. Król, Category Theory and Philosophy, in this volume
18. S.S. Kutateladze, Saunders Mac Lane, the Knight of Mathematics. Sci. Math. Jpn. **63**(1), 4–8 (2005). arXiv:math/0507203
19. E. Landry (ed.), *Categories for the Working Philosopher* (Oxford University Press, Oxford, 2017)
20. W. Lawvere, Adjointness in Foundations. Dialectica **23**(3–4), 28–296 (1969). (see also Reprints in Theory and Applications of Categories. No. **16**, 1–16 (2006))
21. W. Lawvere, Diagonal arguments and cartesian closed categories. Theory Appl. Categ. **15**, 1–13 (2006)
22. H. Leitgeb, Scientific philosophy, mathematical philosophy. Metaphilosophy **44**, 267–275 (2013)
23. J. Lubacz, The Application of Category Theory to Epistemic and Poietic Processes, in this volume
24. S. Mac Lane, *Mathematics: Form and Function* (Springer, Berlin, 1986)
25. S. Mac Lane, The protean character of mathematics, in *The Space of Mathematics. Philosophical, Epistemological, and Historical Explorations*, ed. by J. Echeverria, A. Ibarra, T. Mormann (Walter De Gruyter, Berlin, 1992), pp. 3–13
26. S. Mac Lane, Concepts and Categories in Perspective, in *A Century of Mathematics in America, Part I, volume 1 of History of Mathematics*, ed. by P. Duren, R. Askey, U. Merzbach (American Mathematical Society, Providence, 1988), pp. 323–365
27. S. Mac Lane, *Categories for the Working Mathematician* (Springer, Berlin, 1998)
28. S. Mac Lane, *A Mathematical Autobiography* (A K Peters, Wellesley, 2005)
29. B. Mazur, When is one thing equal to some other thing? in *Proof and Other Dilemmas: Mathematics and Philosophy*, ed. by B. Gold, R. Simons (Spectrum, 2008), pp. 221–241
30. C. McLarty, The roles of set theories in mathematics, in: [19] pp. 1–17
31. T. Leinster, *Basic Category Theory, Cambridge Studies in Advanced Mathematics*, vol. 143 (Cambridge University Press, Cambridge, 2014), also available as: arXiv:1612.09375v1
32. P. Mancosu, Mathematical explanation: problems and prospects. Topoi **20**, 97–117 (2001)
33. S. Popescu, D. Rohrlich, Quantum nonlocality as an axiom. Found. Phys. **24**(3), 379–385 (1994)
34. M.D. Resnik, Structural Relativity. Philosophia Mathematica **4**, 83 (1996)
35. W. Sellars, Philosophy and the scientific image of man, in *Science, Perception and Reality* (Routledge & Kegan Paul, Abingdon-on-Thames, 1963)
36. Z. Semadeni, A. Wiweger, *Wstęp do teorii kategorii i funktorów* (Polskie Wydawnictwo Naukowe, Warszawa, 1972)
37. Z. Semadeni, Asymmetry of Cantorian Mathematics from a Categorial Standpoint: Is It Related to the Direction of Time?, in this volume
38. M. Shulman, Homotopy TypeTheory: A Synthetic Approach to Higher Equalities, in: [19], pp. 36–57
39. B. Skowron, W. Kubiś, Negating as turning upside down. Stud. Log., Gramm. Rhetor. **54**(67), 115–129 (2018)
40. B. Skowron, Using Mathematical Modeling as an Example of Qualitative Reasoning in Metaphysics. A Note on a Defense of the Theory of Ideas. Ann. Comput. Sci. Inf. Syst. **7**, 65–68 (2015). https://doi.org/10.15439/978-83-60810-78-1
41. D. Spivak, Categories as mathematical models, in: [19], pp. 381–401
42. Univalent Foundations Program, *Homotopy Type Theory: Univalent Foundations of Mathematics*, 1st edn (2003). http://homotopytypetheory.org/book/
43. S. Vickers, *Topology via logic* (Cambridge University Press, Cambridge, 1996)
44. K. Wójtowicz, Are There Category-Theoretical Explanations of Physical Phenomena?, in this volume
45. N.S. Yanofsky, A Universal Approach to Self-Referential Paradoxes, Incompleteness and Fixed Points. Bull. Symb. Log. **9**(3), 362–386 (2003)

Chapter 2
Category Theory and Philosophy

Zbigniew Król

Abstract This paper considers the role that category theory can play in philosophy. Category theory is a source of problems, methods and inspiration when it comes to considering both some new and some longstanding philosophical issues. Among the former, the paper draws attention to the ontological interaction between categories and sets, as well as the quantificational criterion of being—to mention just two. Among the latter, it highlights the problem of cognitive access to mathematical objects, and that of the way in which such objects exist. In the context of the development and the ontology of mathematics, I argue in favour of the thesis that category theory is the most "platonic" theory in mathematics. I also point out that category theory impacts significantly upon many standard philosophical positions, providing many counter-examples to popular, often repeated, yet unjustified philosophical claims. The influence of category theory on the foundations and ontology of mathematics is also briefly explored here.

2.1 General Remarks Concerning Category Theory (CT)

First of all, CT proposes not just a new theory or domain of mathematics: it furnishes a new conceptual framework for it. The tools of CT make it possible to speak about and study all (?) kinds of formal mathematical theory, because of the fact that CT is also a foundation for mathematics. Any (?) mathematical object can be investigated with the use of CT, while it is also a fundamental mathematical theory, in that it enables not only the translation of each and every (?) theory into CT-form but also makes it possible to investigate every (?) kind of mathematical object with the use of the CT conceptual apparatus. Moreover, CT enables one to study mutual relations within the network of mathematical theories, or even of mathematical domains. This last remark concerns the *same kind* of mathematical object (or theory), such as the

Z. Król (✉)
International Center for Formal Ontology, Faculty of Administration and Social Sciences, Warsaw University of Technology, Warszawa, Poland
e-mail: zbigkrol@wp.pl

M. Kuś and B. Skowron (eds.), *Category Theory in Physics, Mathematics, and Philosophy*, Springer Proceedings in Physics 235,
https://doi.org/10.1007/978-3-030-30896-4_2

category of all small (or large) groups, sets, algebraic structures or models of theories, as well as *different kinds*, such as the category of all categories or connections between categories consisting of different objects.

One can describe Set Theory (ST) in a similar way. Obviously, CT and ST are not singular formal theories, but rather open domains accompanied by the relevant methods and styles of consideration, together with some basic concepts which can be investigated within many different formal theories. CT would only be a new language for all (?) of mathematics if every ST-description of an object had an equivalent CT-description, and *vice versa*. However, the sentence above is not true, because one can construct mathematical situations where the objects have some different properties that are not discernible from the CT-point of view. The same holds for some (not only ST-) extensional theories and some intensional properties.

To almost every statement concerning the hypothetical properties of CT and its role in mathematics there corresponds a very strong meta-theorem (cf. the examples mentioned above). At the same time, in my opinion, these theorems are not true *in general*—thus I have placed a question mark in some places above. They are only true in some situations and given some assumptions, because it is possible to construct counter-examples to the unlimited versions of these theorems.

ST and CT are affected by much the same kinds of well-known problem and antinomy. For example, is everything a set, or is every mathematical object a set with some additional structure? The answer, moreover, is the same in ST as in CT: i.e. "no". For instance, it is possible to express in ST such sentences as "x is not a set" or "x is not an element of any set and x has no elements". It is also clear that it is something different to speak about one ST-theory, such as a theory of the foundations of mathematics, than it is to talk about multiple different ST-theories in which locally mathematical objects can be "immersed". In my opinion, it is not even the case that "all (*a priori* possible) mathematics" is locally an instance of ST. Many counterexamples can be constructed with the use of alter-theories to different ST- and CT-theories. The *alter*-theory to a given axiomatic theory, whose axioms are logically independent, is a theory whose axioms are negations of the original axioms. In every theory based on an independent and consistent set of axioms, one can replace any axiom by its negation and the resulting theory will also be consistent. In the case of first-order theories such as ZF(C), PA, or the elementary theory of the category of sets (ETCS), there exists an *alter*-theory with many interesting properties and models.[1] *Alter*-ZFC cannot have models describable in ZFC, if both theories are consistent. Thus, if every object is locally describable in ZFC (a CT-theory, …), what are the objects of *alter*-ZFC (*alter*-CT-theory, …)?

Hence, it would be far better to speak about ST- and CT-reductions of mathematics, or ST- and CT-interpretations of mathematics, that are limited in their power of expression. Moreover, such interpretations are not *conservative* in mathematical content. For instance, it is not *exactly* or *ideally* the same to speak about such fundamental mathematical objects as, for instance, natural numbers, in ST, CT or Peano Arithmetic (PA). In different CT-structures, e.g. in the different kinds of topoi,

[1] However, some *alter*-theories of sophisticated theories are trivial.

natural numbers exhibit some standard as well as non-standard properties, including many unexpected or exotic properties. Set theoretical definitions of natural numbers treat (i.e. interpret) those numbers as sets. Thus, within, say, ZF(C), one can ask "if 2 is an element of 4", which is not a property of numbers from the intuitive point of view, and is meaningless in PA. Therefore, each interpretation extends and changes the interpreted concepts, and adds something new to them. PA seems to be closer in that respect to our intuitions. In general, there is a philosophical question to be asked here: do such objects exist independently of the theories known to us—or, if/when the theories contain non-equivalent descriptions of the objects, which theory (and language or conceptual frame) is the *real* and *proper* means for describing *the* numbers and other mathematical objects? Put another way: has any mathematical theory been *discerned* as describing mathematical objects and truths? Such a theory would be an ontologically discerned theory, in the sense that it would be the basic theory for ontological considerations. ST and CT do not themselves furnish such *absolute* mathematical means and theories.

There are some advantages of CT in comparison to ST. For instance, in CT it is easy to study the interrelations between different mathematical theories and domains. Mathematics, from this point of view, is not based on one fundamental theory, but seems to be an expanding network of different structures and corresponding theories which can be locally inquired into along the lines furnished by CT.

2.2 CT and Some Classical Problems in Philosophy

My intention here is to keep matters brief by confining myself to some general remarks, highlighting just some selected *general* philosophical problems whose analysis CT may prove relevant and useful to. Thus, rather than seeking to cover the numerous problems arising in the philosophy of mathematics or logic as they relate to CT, I shall just focus on addressing certain longstanding problems in philosophy from this perspective.

Mathematics, from its early beginnings, has been the main source of examples and archetypical patterns for some of the solutions proposed in general philosophy. Philosophy has itself been closely connected with mathematics, with the development of mathematics having had a major impact on the development of philosophy and *vice versa*: consider, for instance, the philosophies of Plato, Aristotle, Neoplatonism, Descartes, Newton, Berkeley, Hume, Locke, Kant, Mill, Frege, and others. Thus, mathematics shows up at the very centre of philosophical considerations, and both its actual use and our reflections upon it lie right at the heart of philosophy.

The emergence of new mathematical theories such as ST, of formalized mathematical and other languages, and of non-Euclidean geometries, have all exerted a significant influence upon philosophy. On the other hand, both old and new tendencies in philosophy—such as platonism, phenomenology and intuitionism—have influenced pure mathematics. From this point of view, it seems interesting to take into consideration the possible impact of CT on philosophy.

First of all, there are many works in CT which are philosophically interesting *per se*—such as, for instance, the works of William Lawvere concerning categorical set theory. On the other hand, many classical problems in philosophy can be analyzed in the context of formal ontology using mathematics, logic and formal methods. Therefore, CT can be seen as a new method in formal ontology and philosophy, and especially in analytical philosophy. Some of the deep-level results emerging on a CT-based treatment differ from what they would be on an ST-approach. CT is not merely a new language, useful for describing the same problems or objects that can be expressed when we employ ST-based tools. This last fact is a consequence of many general properties of CT: such as, for instance, the abovementioned, non-conservative description of mathematical objects within different theories and structures, or the use of intuitionistic logic. There is, then, most certainly room for the application of CT in philosophy and formal ontology. However, such an approach is not yet especially popular amongst philosophers.[2]

2.2.1 CT and Ontology

(a) One of the most important ontological problems is the topic addressed in the debate concerning ontological monism versus pluralism. Are there only sensual or physical (material) objects? Is there only one way of being: i.e. do only material or real objects exist? CT can enhance our discussions about, and possibly even help solve, many such questions. However, at the same time this is so in a very limited way: it is only useful in the construction of some counterexamples to certain particular standpoints. For instance, the central position in analytical philosophy keeps Quine's quantificational criterion of existence (QCE), albeit with some variations. Quine argued for monism. There have been some experiments introducing theories with multiple different existential quantifiers that can correspond to certain different ways of being. Such theories can be discussed with the use of CT—cf. [9]. The analysis demonstrates that different quantifiers, as well as QCE, are irrelevant in ontology and rather give rise to technical and logical problems on their own account—ones which are certainly most interesting from a mathematical point of view, but which cannot be said to enrich pure ontology. Below are two examples illustrating the possibility of constructing CT-counterexamples to QCE:
(1') There are categories in which quantifiers are not interpretable, e.g. categories of cartesian closed categories.
(2') In the case of some formulae in the Zariski topos, one encounters the following: $\exists x. \sim\sim F(x)$, As is well-known, $F(x)$ states the existence of some invertible infinitesimals. However, one cannot infer from the above that $\exists x.F(x)$ in this topos. Thus, we are faced with the situation that some so-called *inhabitant individuals* (which are accessible after the application of forcing in the Basel topos)

[2]See the various works listed in the bibliography: for instance, [11].

can be added, but it is impossible to express this fact against the background of QCE. This leads to the construction of a Basel topos in which one can express this fact—i.e. that $\exists x.F(x)$—cf. [16, pp. 285–292, 65].

(b) The above-mentioned modern questions are connected with the old problem of platonism in philosophy and mathematics. The development of mathematics made it possible to try to describe and answer many questions, such as what *kind of beings* are mathematical objects? The original answer to the first of those questions was the one given by the Pythagoreans: i.e. that "everything is a number", and because "number" for the Greeks only meant "natural number" ("1" was not a number, but the principle of the generation of numbers, and they did not know "0"), some problems soon emerged. If everything in the world is a number, then especially everything in mathematics must be a number, and especially every geometrical object will be a number. However, the discovery of incommensurability demonstrates that if the side of a square is a number "1", the diagonal cannot be a number, because this number would have to be even and odd simultaneously, which is impossible. Therefore, the ancient mathematicians tried to construct a single mathematical theory in which everything is a natural number or is describable with the use of natural numbers. They considered many variants: for example, the introduction of more than one series of natural numbers, each with its own unit "1", incommensurable with other "1's". Theaetetus of Athens demonstrated that it is not possible to describe the totality of lines (finite sections) in geometry with the use of a finite number of "copies" of natural numbers [cf. Book X of Euclid's *Elements*, and [8, Chap. 4 and 5]]. Modern versions of the above problem take the form of such putative answers as "everything is a set", or "everything is a category", both of which have their own well-known difficulties, such as the ST- and CT-antinomies. Such answers are rather restrictive from the point of view of pure mathematics and mathematical practice. They are also irrelevant from the point of view of the history of mathematics. For instance, it is quite artificial to analyze ancient mathematics with the use of modern algebra, ST, or CT. Moreover, the application of modern tools to ancient mathematics in some cases renders the (ancient) problems under consideration more difficult and complicated than they were on the original (ancient) approach. One such example is the problem of change of the basic line in the classification of incommensurable lines in Book X of *Elements* [8, Chap. 4]. The modern notation is ineffective and too complicated when it comes to the analysis of this problem. ST- and CT-reductions of mathematics seem to indicate that the ancient mathematicians also took into consideration some sets or categories, albeit subconsciously and without any explicit mention of the fact that they were doing so. Historical analyses indicate that the objects they were engaged in considering were, intensionally, entirely different from sets or categories. For instance, the first explicit treatment of geometrical objects as sets of points originated in 1804 with Bolzano [2]. Even the well-known treatment of a (real) line as a real axis only became possible after Cantor's treatment of this problem. Hence, it would be better to speak about ST- and CT-interpretations of the mathematical universe than about "reductions". What I mean is that such interpretations cannot resolve the problem of what the subject matter of mathematics is and what mathematical objects are from an ontological point of view. The most relevant answer to these questions is that there are many different objects or universes of mathematical discourse and we can grasp the mutual relations between them. Given such a goal and approach, CT seems highly appropriate.

Coming back to platonism, CT cannot give us more information about the way in which mathematical objects exist. On the other hand, it can show that there are many mutually irreducible mathematical objects, even from the point of view of extensional theories.[3]

(c) Turning to the next point in this review of the potential impact of CT in the field of ontology, it is necessary to indicate some ontological consequences stemming from the sheer fact of the internal ontological properties of CT. CT is the most "platonic" theory (or way of thinking) that there is in mathematics, not just ahistorically speaking, but also from a historical point of view. Mathematical platonism is usually considered to be the conviction that there are—or the acceptance of the existence of—ideal, atemporal, eternal, non-physical, unchanging objects (beings), where these are taken to be the subject matter of mathematics. Such a conviction, to be sure, can be explicit or implicit. In cases where such convictions are explicit, they would appear to be external to mathematics and inessential to the creation of mathematical knowledge itself. The most important are those implicit platonic convictions which manifest themselves in determinate methods of mathematical inquiry, or in some special way of being of the subject who creates mathematics. A well-known example of such implicit platonic convictions is given in the form of the problem of the decimal expansion of the number π: is there a sequence "0123456789"? If somebody thinks that the answer is either "yes" or "no", he or she holds an implicitly platonic commitment. Behind this kind of platonism lie various strictly determined methods, such as the use of the law of the excluded middle, indirect proofs, classical logic, etc. Some of those methods are "less mathematical"—such as, for instance, the possibility of "seeing a given theory from the outside", or the appeal to mathematical intuition. The reader can find more examples in [6], as well as in existing literature [1]. From this point of view, CT is even more platonic (non-constructive) than ZFC, NBG, etc. It makes use, in an almost unlimited way, of the possibility of "seeing" theories and objects "from the outside": each and every category of object is set up as consisting of objects which, at least usually, are developed mathematical structures of some sort—e.g., groups, algebras—and defines external relations involving them. CT, then, treats mathematical objects as already existing, and ready-to-use: "Let Ens be *the category of all sets*, Set *that of all small sets*, Mon *that of all small monoids*, Cat *that of all small categories*, ..." At the same time, one can consider functors, arrows, etc. between everything—large or small categories. CT uses non-constructive totalities of objects, and it is its basic method to assume the existence of such totalities and discern the relations between them.

[3] I explain more precisely what "extensional theories" are in [7, 8].

Such a possibility enables the use of mathematical intuition—see the material devoted to epistemology below.

On the other hand, the internal logic, in the most important cases, is intuitionistic. Topoi furnish models for intuitionistic logic of any order. CT demonstrates the fruitfulness and effectiveness of this sort of platonic approach, in which one can inquire into limits affecting the use of different kinds of logic in mathematics. More specifically, CT enables the study of limits that bear on platonism and platonic methods.

CT allows us to study platonism in physics. We have already said that CT treats mathematical objects as already existing and ready-to-use. In the same way, in physics, mathematical objects are ready-to-use *in the absence of* formal theories. For instance, natural numbers are "in use" in quantum physics (the numbering of quantum states) as if they were unique—i.e. independent of the models provided by certain formal theories. Therefore, we have a model-theoretic approach to quantum gravity which is similar to, or "translatable" into, topos-theoretic language and a topos-theoretic approach (Isham).[4]

In brief, then, CT cannot add anything more to our purely ontological considerations concerning the mode of existence of mathematics and mathematical objects. The objects of the CT-universe are, in the same way, *abstract* or *ideal* or *non-physical*—as objects of set theory or algebraic structures, or the objects considered in ancient mathematics (squares, numbers, triangles, lines, circles, etc.). However, CT falsifies some "ontological" views, such as set-theoretic platonism (i.e. the thesis that "everything is a set" and, consequent upon this, the claim that the question "What exists in mathematics?" has an answer independent of the content of the/a universe of sets. The foundations of mathematics can be interpreted ontologically as an experiment of sorts, aimed at determining what the basic concepts and structures—i.e. beings—in mathematics are. However, there is an infinite number of possible foundations of mathematics, and from that fact follows the impossibility of using CT to define the ontological content of the mathematical universe.

2.2.2 CT and Epistemology

In my view, CT points towards the possibility of inquiring into the same mathematical objects from radically different perspectives, appealing to very different intuitions. Thus, CT adds something essential to classical, purely epistemological inquiries dealing with our epistemic access to mathematical reality. However, given that it concerns our epistemic access to the different ways of being of the subject matter of mathematics, we should recall that the objects of the CT-universe are *abstract*, *ideal* or *non-physical* in much the same way as the objects of set theory and algebraic structures, or those considered in ancient mathematics (squares, numbers, triangles, lines, circles, etc.), are. Therefore, both the objects of CT and CT itself are given *to the*

[4] Cf., for instance, [3–5, 10].

conscious subject in the same *type* of phenomenological act as other mathematical or abstract objects. These objects, then, are intentional objects.[5]

At the same time, one can consider whether CT-objects are given in the same kind of conscious act as sets, algebraic objects, and so on: i.e. whether there are epistemic and phenomenological differences in respect of our conscious access to different types of object in mathematics. Nobody has yet considered the above question. The results and theoretical apparatus elaborated upon in Husserl's *Logical Investigations* can be useful here. Thus, in epistemology, CT can be called upon mainly as a source of examples in philosophical case studies. Meanwhile, it is possible to model and study many epistemological problems in formal ontology and CT. Problems pertaining to the limits of the use of language in philosophy and, especially, in epistemology, which can be treated with the use of CT and other formal methods, belong to this problem set. Nevertheless, in my opinion, such an approach is rather external to epistemology and its essential methods and problems.

The next group of examples of questions which can be investigated with the use of CT are as follows:

1. Is the structure of reality the same (isomorphic, "similar", etc.) to the structure of a language or a theory?
2. Is the structure (subject) of mathematics the same or "similar", or does it correspond to, the structure of the language in which one conducts mathematics?
3. Is there any form or subject matter of mathematics that is given to us without involving any sort of language?

What is mathematical intuition? How does it operate? And how is epistemic access to mathematical reality possible? These are major questions, not only in the philosophy of mathematics but also in general epistemology. The best way to approach them is via phenomenology and hermeneutics. Nevertheless, the possibility of discovering something in mathematics in a non-formal or intuitive way rests on the possibility of making use of some objects from formal theories in an informal way. One can elaborate some unspecified "points" taken from a model of a formal theory and *imagine* that these are not "points", but rather other full-blooded objects: groups, algebras, or categories. For instance, it is possible to imagine that, in a real line, the initial "points" are substituted by some groups, linear spaces or structures. Thus, it is possible to investigate some objects in which some other objects are immersed or have been substituted, which are therefore *things* which possess some properties inexpressible in the initial theories. There are plenty of kinds of such substitutional theories and "substitutional models". For instance, global homogeneous substitutional models (where one substitutes for every "point", or other objects, in the initial theory the same kind of objects described in other theories), or local homogeneous ones (where one substitutes the same kind of objects only for some chosen "points"), or global and local non-homogeneous substitutional models, as well as self-substitutions (where we use

[5]For Ingarden, mathematical objects are ideas with rigid essences. However, in the opinion of Ingarden, sets are only some sort of purely intentional object, without any such rigid essences. There is a difference between Husserl's and Ingarden's conceptions of intentionality. In the following, I prefer the concept of intentionality given in Husserl's *Logical Investigations*.

in the substitution objects defined in the initial theory—for instance, substituting for every point in a Euclidean plane a straight line). There is a well-defined description of such models in CT. For instance, to every global substitution there corresponds a bundle where the initial set of "points" is a set of indexes and the objects to be substituted are members of a stalk space. Thus, there is a natural topos-structure corresponding to a global substitution. More information about these kinds of model, together with their possible applications and indications concerning CT-formalization, can be found in [9].

2.2.3 CT and Epistemology Contd.: Theories of the Development of Mathematics and Science

CT, conceived and pursued in terms of case studies of various sorts, yields fundamental information concerning both the *a priori* status of mathematics and the laws involved in the development of mathematics. These kinds of results are also important from a general epistemic point of view. Beyond this, CT demonstrates that there are alternative formulations of some basic mathematical theories which start from different intuitions. For instance, first-order descriptions of the Set-category and ZFC are equivalent—the models are, in a sense, equivalent. The ETCS (Lawvere) starts without the concepts of $\in$-relation or element, or of the set as a collection of elements. The above example demonstrates that, from an *a priori* point of view, set theory could have appeared first as an ETCS in the history of mathematic, and that the creation of Cantor's, Dedekind's or Frege's set theories (as well as any other mathematical theory) is determined not only by *a priori* considerations, but also by social, cultural, historical, etc., factors.

Thus, CT demonstrates that mathematics is, in a sense, independent of the intuitions underlying its creation. However, it is impossible to create something in mathematics that is entirely without any intuitive content or convictions. Thus, CT could furnish many case studies relevant to the problem of what the connection might be linking mathematics with primordial intuitions. Again, in my opinion, the basic tools needed are those provided by the methods and philosophical approach presented in Husserl's *Logical Investigations* (*passim*).

The next example of such a case study would be this: that it is possible to demonstrate that ancient geometry, as known from Euclid's *Elements*, has an alternative formulation based on different intuitions. Every theorem from Book I of *Elements* can be obtained without the intuitions concerning points, lines, circles, triangles, etc.—cf. [8].

The final possibility I wish to touch upon here is that of studying the logical structure of the brain (or, if the reader prefers, the mind) and its neural activity, insofar as these can be effectively modelled in CT. Thus, CT could prove to be a useful tool in cognitive studies, and in naturalistic and causality-based approaches to epistemological matters.[6]

2.3 Conclusion

Our most important conclusion here is that CT cannot replace purely philosophical methods in ontology and epistemology—especially phenomenological ones. Nevertheless, it is an extremely useful and effective tool in formal ontology—one irreducible to other formal tools, such as ST and its language and methods. It is a great source of counterexamples which can falsify many standpoints in philosophy—especially in ontology and analytical philosophy. The initial emergence of CT, together with its subsequent development and achievements, are a rich source of possible case studies that have the potential to cast light on a number of philosophical problems, not only in pure ontology and epistemology, but also pertaining to the laws involved in the development of mathematics, knowledge and science in the context of the use of mathematical intuition. In my view, then, it is impossible to work competently with many problems in the philosophy of science, the philosophy of mathematics, or analytical philosophy—especially where the ontology and epistemology of mathematics are concerned—without having in one's possession at least an elementary knowledge of the methods and accomplishments of CT.

Appendix: The Commentary of an Anonymous Reviewer

The paper deals with very important and so far only partially explored topic of relations between category theory (CT) and philosophy. There are (as far as I know) not many works in this field. On the other side, many existing mathematical papers on CT contain strongly philosophical parts (e.g. Lawvere). Nevertheless, there is a shortage of systematic studies of CT and philosophy. The Author presents several examples, both in philosophy and mathematics, where such studies could be performed.

In particular I am aware of the Author's work on alter theories which are mentioned in this paper. Let me comment on this point more carefully. In general an alter-theory is built as an associated theory to any existing axiomatic theory with recursive set of axioms, consistent and logically independent. Then one can negate one or more of these axioms which leads to the variety of possible alter theories. It is quite surprising that such set of alter-theories can be used as a tool for exploring the original theory. I think that this is also very nontrivial construction regarding existence in mathematics and possibly in philosophy. This also sheds some light on a would-be formal theory for ontology (at least in mathematics). To explain these points let me consider a model M of ZF(C). Then one can modify slightly the procedure

[6] Cf. the possibilities opened in [12–14]. See also [15, 17, 18].

of building an alter-theory: It relies rather on creating both the (model of a) theory and the (model of the) alter-theory. Historically such an approach has been realized by forcing extensions of M. There typically appear independent sentences on ZF(C) in the extensions, e.g. continuum hypothesis in one extension and its negation in the other. One extension can be seen as the model for a theory and the other for the alter-theory. One can ask the question: Is there anything fundamental for mathematics in the entire procedure of creating the pairs (Theory, Alter-theory) as above? Recent results of Joel Hamkins (e.g. [20]) show that the entire structure of the set of such forcing extensions (partial order) is a fundamental entity in mathematics, called the multiverse. The best recognized multiverse is the one where the initial model M is countable transitive and the forcing extensions are just Cohen extensions. The true challenge would be to recognize the multiverse starting from the constructible universe $\mathbf{L}$. But the ultimate goal would be to investigate the multiverse corresponding to the universe V of sets of von Neumann. By now it remains among very ambitious, though beyond current possibilities, aims of the approach. In this way the procedure of passing to the pairs (Theory, Alter-theory) by means of forcing extensions appears as fundamental in foundations of mathematics. Finally, the categorification of the multiverse is possible and would lead to another interesting connections between CT, set theory and foundations of mathematics.

Regarding existence in mathematics let us make a simple observation: Given ZF(C) model M we can add extra axioms which leads to the more powerful capabilities for proving the existence of certain objects. The profound example is that of the large cardinal axioms added to ZF(C). Then one can prove the existence of things such as $0^{\#}$. They correspond to real numbers which are not present in any forcing extension of M but exist in $\mathbb{R}$ in V under suitable large cardinal hypothesis (LC). Without LC, $0^{\#}$ does not exist. Another example is the constructible axiom $V = \mathbf{L}$ added to ZFC. In such models, nonconstructible reals certainly do not exist and their existence relates again to LC when $\mathbf{L} \subset V$ (under $\mathbf{L} \neq V$).

The question is whether there are any fundamental model M and 'the most fundamental' LC axioms which would be 'the best' for the entire mathematics. If they exist, this could shed light on a would-be formal theory for ontology in mathematics. As argued by Hugh Woodin, one candidate for such a formal theory could be certain extension of $\mathbf{L}$ by all real numbers, i.e. $\mathbf{L}(\mathbb{R})$; then, LC is given by a suitable set of Woodin cardinals (see e.g. [21]). The point is that $\mathbf{L}(\mathbb{R})$ is stable under forcing extensions and it survives the modifications. This seems to be a very fundamental approach in mathematics, which leaves space also for categorical reformulation (cf. [19]). In all cases discussed above, an important thing was the behaviour of the models under forcing extensions. They can be seen alternatively as certain instances of alter-theories. However, building alter-theories seems to be a wider and simpler procedure than forcing extensions and consequently a less specific one. However, it is quite interesting that such simple point of view exists at all, referring directly to deep questions in mathematics.

Even though the above remarks are of rather purely mathematical nature, they carry also potential for philosophy. However, I am not competent enough to decide precisely this case. Similarly, I do not know whether these remarks should influence the current chapter or should be considered rather in a separate work. I will leave this issue to the Author's judgement.

References

1. P. Bernays, Sur le platonisme dans les mathématiques. L'Enseignement Math. **34**, 52–69 (1935)
2. B. Bolzano, *Betrachtungen über einige Gegenstände der Elementargeometrie* (Karl Barth, Prague, 1804)
3. A. Döring, C.J. Isham, A topos foundation for theories of physics: Ii. daseinisation and the liberation of quantum theory. J. Math. Phys. **49**(5), 053516 (2008)
4. A. Döring, C.J. Isham, A topos foundation for theories of physics: I. formal languages for physics. J. Math. Phys. **49**(5), 053515 (2008)
5. J. Król, *Model-theoretical approach to quantum gravity*. Ph.D. thesis, the Institute of Physics at the University of Silesia, Katowice (2005)
6. Z. Król, *Platonizm Matematyczny i Hermeneutyka* (Wyd. IFiS PAN, Warszawa, 2006)
7. Z. Król, Uwagi o stylu historycznym matematyki i rozwoju matematyki, in *Światy matematyki, Tworzenie czy odkrywanie?*, ed. by I. Bondecka-Krzykowska, J. Pogonowski (Poznań, Wydawnictwo Naukowe UAM, 2010), pp. 203–234
8. Z. Król, *Platonism and the Development of Mathematics: Infinity and Geometry* (Wyd. IFiS PAN, Warszawa, 2015)
9. Z. Król, J. Lubacz, What do we need many existential quantifiers for? Some remarks concerning existential quantification, monism and ontological pluralism. manuscript
10. J. Król, Exotic smoothness and noncommutative spaces. The model-theoretical approach. Found. Phys. **34**(5), 843–869 (2004)
11. E. Landry, *Categories for the Working Philosopher* (Oxford University Press, Oxford, 2017)
12. W.S. McCulloch, Machines that think and want, in *Brain and behavior: A Symposium. Comparative Psychology Monographs, vol. 20, chapter 1*, ed. by W.C. Halstead (University of California Press, Berkeley, 1950), pp. 39–50
13. W.S. McCulloch, Agathe tyche of nervous nets—the lucky reckoners, in *National Physical Laboratory Symposium*, chapter 2, vol. 10 (1959), pp. 613–625
14. W.S. McCulloch, W. Pitts, A logical calculus of the ideas immanent in nervous activity. Bull. Math. Biophys. **5**(4), 115–133 (1943)
15. J.M. Mira, Symbols versus connections: 50 years of artificial intelligence. Neurocomputing **71**(4–6), 671–680 (2008)
16. I. Moerdijk, G.E. Reyes, *Models for Smooth Infinitesimal Analysis* (Springer, New York, 1991)
17. D.H. Perkel, Logical neurons: the enigmatic legacy of warren mcculloch. Trends Neurosci. **11**(1), 9–12 (1988)
18. H. Von Foerster, *Computation in neural nets, in Understanding Understanding* (Springer, New York, 2003), pp. 21–100
19. J. Adamek, J. Rosicky, *Locally Presentable and Accessible Categories*. London Mathematical Society Lecture Notes Series, vol. 189 (Cambridge, 1994)
20. J.D. Hamkins, The set-theoretic multiverse. Rev. Symb. Logic. **5**, 416–449 (2012)
21. W.H. Woodin, *The Axiom of Determinacy, Forcing Axioms, and the Nonstationary Ideal*, 2nd edn. (De Gruyter, Berlin, 2010)

Chapter 3
Are There Category-Theoretical Explanations of Physical Phenomena?

Krzysztof Wójtowicz

Abstract The problem of mathematical explanations in science has been discussed extensively in the philosophical literature in recent years. This paper is devoted to the question of whether category theory can offer explanations of physical phenomena. My claim is that it cannot. Nevertheless, it cannot be denied that CT contributes to our understanding of physics, therefore in this paper I discuss whether this fact can be accounted for in terms of explanatory value. I argue that investigating the (possible) explanatory virtues of CT is important for the discussion concerning mathematical explanations on an abstract level. So, it contributes not only to understanding the role of CT in physics, but also to elucidation of the notion of mathematical explanation in science.

3.1 Introduction

Category theory (I will use the acronym CT throughout the text) is a well-established branch of mathematics that impacts many mathematical disciplines (e.g. differential geometry, algebraic topology or computer science), and also physics (these applications are covered elsewhere in this volume). Enthusiasts of CT claim that it provides sound foundations for mathematics that are different from the standard set-theoretic foundations. CT has also an important impact on the philosophy of mathematics as many authors claim it to be the proper expression of structuralistic intuitions concerning mathematical ontology. The philosophical impact of category theory is not restricted to the philosophy of mathematics as it provides important conceptual tools with which to discuss fundamental metaphysical concepts such as identity or analogy.

One of the fundamental philosophical questions concerning mathematics is its role in scientific explanations. This topic has been widely discussed in recent years, and the aim of this paper is to examine the status of CT in this respect. The problem

K. Wójtowicz (✉)
Department of Logic, Institute of Philosophy, Warsaw University, Warsaw, Poland
e-mail: kwojtowi@uw.edu.pl

M. Kuś and B. Skowron (eds.), *Category Theory in Physics, Mathematics, and Philosophy*, Springer Proceedings in Physics 235,
https://doi.org/10.1007/978-3-030-30896-4_3

of possible category-theoretic explanations in science is a special case of the more general problem of the explanatory role of mathematics in science. This is a fiercely debated topic; an example of an important contribution is [15], which addresses the problem of whether there are genuine mathematical explanations in science, i.e. whether mathematics *an sich* has explanatory power.[1] At first sight the answer seems obvious as mathematics is crucial for modern science and we cannot even imagine physics (chemistry, biology, ecology, economics...) without it. But the problem is more subtle: the omnipresence of mathematics in science does not entail that it plays an explanatory role. We might think of mathematics as of an auxiliary system that is devoid of content, or we might conceive it just as a convenient (perhaps even necessary) way of representing ("indexing") phenomena, but without inherent explanatory virtues.

The discussion is not settled, but in this paper I assume that the notion of mathematical explanation is sound: (at least some) mathematical notions, techniques and disciplines can provide explanations of physical phenomena, but it is clear that mathematical disciplines profoundly differ in their potential applicability. Obviously, differential equations are much more likely to do some explanatory job in physics than set theory, which deals with (for example) the relative consistency of large cardinal axioms, or embeddings of inner models. A natural question is whether category theory also plays a part in this explanatory work, i.e. does it offer any kind of explanations in science?

I think that it is not explanatory in the sense which seems to be standard in the philosophical literature; however, this negative claim is a good starting point for the discussion concerning the epistemic contributions of category theory, in particular its (perhaps nonstandard) explanatory role. This statement might be re-interpreted in a "dual form" as a claim concerning the notion of explanation.[2] This issue really involves two sub-problems: (1) What is the role of CT in physics? (2) How do we interpret the notion of "mathematical explanation in science"? These two questions might be viewed as a special case of more general problems: (1) Does theory T have any explanatory contributions for (within) science S? (2) What do we mean by "the explanatory contribution of T within S"?

So, analyzing the (possible) explanatory role of CT also contributes to our understanding of the notion of explanation. Indeed, CT shows its power in the context of fundamental issues that do not represent concrete phenomena and which should perhaps be considered metaphysical rather than physical—its strength is revealed on a deep conceptual level.[3]

In part (1) (*Mathematical explanations of physical phenomena*) the problem of the explanatory role of mathematics in science is discussed. I present some classic

[1] An up-to-date presentation with an extensive bibliography is [19].

[2] So, this "dual form" is something like this: assuming CT has no explanatory force, what does this fact tell us about the notion of explanation?

[3] CT might be helpful in elucidating philosophical notions, perhaps even by providing formal explications. These investigations exceed the scope of this article (some of the issues are hinted at in [13]).

examples; in particular, I mention topological explanation as an important case that is a "contrast class" for the hypothetical category-theoretical explanations. I argue for the thesis that no category-theoretic explanations in physics can be found—at least not in the sense which seems to be standard in the literature on mathematical explanations in physics. So, the role of CT can be claimed to be explanatory only if the notion of explanation is understood in a different way.

In part (2) (*Category theory in physics*), I discuss (as an example) Isham's program of Topos Quantum Theory (TQT) and claim that it shows that the epistemic virtues of category theory are located on the metalevel: it might therefore be called "a theory of theories" [25, p. 400] rather than a theory about a particular subject. I also comment on the problem of whether gaining metatheoretical knowledge can contribute to our understanding on the object level. Finally, I argue that the explanatory contributions of CT have a philosophical rather than a specific scientific character.

A short conclusion (part 3) concludes the text.

3.2 Mathematical Explanations of Physical Phenomena

Recently, the problem of the explanatory role of mathematics within physics (and science in general) has received much attention.[4] The question is whether there are any mathematical explanations in science, i.e. whether it is mathematics which provides explanatory power. This is a subtle problem as we have to distinguish between scientific explanations that merely make use of mathematics, and explanations where mathematics does the genuine explanatory work.

Perhaps mathematics cannot by definition explain anything as it only serves as a tool for capturing and expressing relationships between physical phenomena without making any genuine contribution—no "added value". The use of mathematics in such procedures might even be necessary, but it is rather a kind of map with a merely representative role. The problem is intricate: are there *physical* properties of *physical objects* which are not expressible in purely physicalist language?[5] According to the doctrine of "abstract expressivism", mathematics has a very special function: it makes it possible to express claims about the physical world which would not be

[4] I refer to the philosophical literature concerning the problem of mathematical explanations of physical phenomena (and mathematical explanations in science in general). Even if there is disagreement concerning the very existence of mathematical explanations, there seems to be a kind of consensus concerning the problem of what they might be (if they existed).

[5] The notion of a purely physicalist language might not appear to be very clear. For instance, we might think even of the language of calculus as physicalist (at least to some extent), as the notion of derivative has a very clear physical inspiration and interpretation. More abstract mathematical theories do not have such direct connections to observations and the physical meaning of terms can be understood only indirectly. The term "nominalistic language" could be used here, as it seems to be better defined in philosophical literature; however, the discussion of this topic is outside the scope of this paper.

expressible otherwise.[6] The literature on the subject is immense, so I am only able to give a few examples without going into details; they vary in their technical intricacy (some being very elementary), but all are meant to illustrate a general phenomenon: when looking for an explanation of some physical fact, we turn our attention to the mathematical structure of the problem rather than to the network of causal relations, the underlying mechanisms, the particular course of events etc. For instance, the fact that the famous Seven Bridges of Königsberg cannot be crossed in such a way that every bridge is crossed exactly once is explained by a simple theorem in graph theory, not by examining the physical properties of the bridges (e.g. the material they are made of, their dimensions, their temperature distribution, the mental processes that accompany the attempts, etc.). Similarly, the structure of honeycombs is explained by a mathematical theorem: hexagonal tiling minimizes the total perimeter length [6], therefore bees do not waste wax and they gain an evolutionary advantage.[7]

An important class that is receiving growing attention is topological explanations, which are found in biology, ecology, medicine, brain sciences and even social science.[8] Some accounts of theories of scientific explanation focus on exhibiting concrete causal dependencies (i.e. identifying the causal nexus of the phenomena), in particular on the mechanisms that underlie particular processes. These theories are contrasted with topological explanations. Kostić gives a clear characteristic: "Very abstractly, the mechanistic explanation describes entities and activities that are organized to produce something or to perform a process or a function. Different levels of organization or properties of the same system can be explained by describing different mechanisms. In contrast, topological explanations explain by a reference to structural or mathematical properties of the system (e.g. graph-theoretical properties, topological features, or properties of mathematical structures in general), and abstract away from the details of particular causal interactions or mechanisms" [12, p. 2].

Topological explanations have some important features which make them especially interesting for our discussion.[9] They are abstract (as they refer to abstract, mathematical properties of the system), but still retain strong and direct links with the physical situation in question. In a sense, they provide a contrast class for the (hypothetic) category-theoretical explanations. Huneman characterizes the general

---

[6] "Numbers enable us to make claims which [...] we [...] would otherwise have trouble putting into words" [28, p. 230]. Of course, what is important here is not numbers as such, as numbers are very simple mathematical objects. What is at play is the usually very abstract mathematical structures that provide the expressive power of the theory: "Let 'abstract expressionism' be the doctrine that mathematics is useful in science because it helps us to say things about concrete objects which it would otherwise be more difficult, or perhaps impossible, for us to say." [16, p. 600].

[7] However, Räz argues (very convincingly), that this example rests on a misunderstanding and offers a different account, albeit in the spirit of mechanistic explanations [23, 24].

[8] Darrason [1] presents many examples of topological explanations within medical genetics and network medicine. Very "down-to-earth" examples (e.g. epidemiology or urban traffic density) and a discussion can be found in [22].

[9] It is disputed whether these explanations are really different from mechanistic ones, but this issue will not be discussed here.

features of topological explanations in the following way (in the context of biology): "a topological explanation is an explanation in which a feature, a trait, a property or an outcome X of a system S is explained by the fact that it possesses specific topological properties" [10, p. 117]. So, we can reformulate the question and ask whether anything similar can be offered by category theory, i.e. can we explain a physical phenomenon by the fact that the system possesses specific category-theoretical properties? In other words, is there some *inherently category-theoretical* theorem which can be applied to physical situations in a way similar to the examples discussed above? It is important to stress that there is a tension between genuinely category-theoretical contributions (which inherently use category-theoretical notions) and classic results that are merely expressed in a different language.[10]

I assume in this paper (as a working hypothesis) that the notion of mathematical explanation in science (physics, biology, chemistry, medicine, cognitive science, etc.) is sound, i.e. that there are genuine mathematical explanations in science. All the examples given above involve quite "down-to-earth" mathematical theorems (and theories). This is also exemplified by topological explanations as many results which are important for topological explanations stem from topological graph theory.[11] In these applications, mathematics often directly represents physical objects (as might happen, for example, in the case of combinatorial problems, e.g. a graph representing the bridges in Königsberg), or describes, for instance, the phase space of possible states or the space of possible trajectories which are linked with a physical system in a natural way.[12] We might say that the level of abstractness is not very high and mathematics is applied to a given physical situation in a rather direct way.[13]

But no examples of an explanatory relation *in this sense* can be given for category theory. We can imagine a situation in which a physicist is helpless in modeling/explaining a physical phenomenon (e.g. heat transport), and a mathematician trained in, for example, stochastic differential equations provides tools to solve the problem in such a way that the model makes it possible to build a more effective engine. But it is very improbable that such direct help could be provided by a category theorist. This is not surprising: CT offers very general (and very subtle) methods for analyzing concepts on a high level of abstractness. So, it is useful for high-level

[10]Marquis discusses this problem in some detail, in particular he observes that CT is often regarded as just "a useful tool to organize, present and develop certain areas of mathematics that are usually considered as being already given in a different setting" [20, p. 250]. Even if such results can be reformulated—"translated" into CT—it does not make them "inherently category-theoretical". (For examples of such reformulations of ordinary mathematical theorems, see for instance [9]).

[11]Some general, "topological" properties of nets are used to explain phenomena in biology or ecology. Nets are a very concrete mathematical object in the sense that we can even point to exemplifications of such structures in our world in a straightforward way (so we can think of a kind of isomorphism). This approach was started by [26].

[12]The particular case of phase spaces is discussed, for instance, in [18].

[13]If we think of the connections between mathematics and the world in a deep sense, Heisenberg's famous quotation "In fact the smallest units of matter are not physical objects in the ordinary sense; they are forms, ideas which can be expressed unambiguously only in mathematical language". From this point of view, the distinction between physical and mathematical objects becomes blurred, and loosely speaking the ultimate structure of reality is the reality of mathematical facts.

(broadly speaking—meta-theoretical) analysis, but not on the level of describing particular physical phenomena (like heat transport in a rod or the shape of honeycomb cells). It is well-suited to analyzing fundamental properties of conceptual systems, theories or models.

The relationship between the notions of explaining and predicting is subtle, and certainly they cannot be identified. But it seems that there is a "minimal predictive input": if a model has no predictive potential at all, it is doubtful whether it really has explanatory character. And in this respect CT has not much to offer when modeling very concrete physical phenomena.

So, my claim is that in this sense there are no category-theoretical explanations of physical phenomena, in spite of there being mathematical explanations. But category theory already has an established position in physics and its role seems to be growing.[14] The claim that CT does not enhance our understanding of science would clearly be false, so the interesting question is what kind of epistemic contribution does CT offer: perhaps it is a different kind of explanation? If so, in which sense does it explain, and what are the *explananda*? Analyzing the role of category theory also contributes to our understanding of the notion of explanation.

3.3 Category Theory in Physics

I focus on a special example of the category-theoretic contribution to physics, namely Topos Quantum Theory.[15] I do not even dare to speculate whether there are any chances for this account to become commonly accepted or whether it will be considered just a theoretical curiosity of no impact on the practice of theoretical physics,[16] but it exemplifies a possible role of CT: I think that it is particularly well-suited for the purpose of analyzing possible epistemic contributions of CT, and in particular to attempts to understand the explanatory role it might play.

Broadly speaking, the aim of TQT is to present a counterpart of the standard version of quantum theory, but couched in category-theoretical terms, in particular using the notions from topos theory. We might say that physical theories are expressed in a different language, this translation being due to the special properties of topoi. The fact that every topos has an internal language is a crucial fact: "Thus constructing a theory of physics is equivalent to finding a suitable translation of the system language, $L(S)$, to the language, $L(\tau)$, of an appropriate topos τ" [2, p. 43].

[14]In [7] we find several articles concerning different category-theoretical ideas in modern physics [14] gives a broad perspective on the applications of category-theoretical notions in different areas of physics, mathematics and philosophy.

[15]This is the program of Isham, Döring, Butterfield and other investigators; see for instance [2, 3] (and subsequent papers); [5].

[16]I think that for some (many?) physicists, investigations of this kind will forever remain "abstract nonsense"—devoid of physical meaning—while for others such investigations lie at the heart of modern physics.

The motivation for developing TQT can be called methodological or even philosophical as the authors argue for a "neo-realist" physics: "So, in toto what we seek is a formalism that is (i) free of prima facie prejudices about the nature of the values of physical quantities—in particular, there should be no fundamental use of real or complex numbers; and (ii) 'realist', in at least the minimal sense that propositions are meaningful and are assigned 'truth values', not just instrumentalist probabilities of what would happen if appropriate measurements are made." [2, p. 11].[17]

One of the reasons that a realist interpretation of quantum mechanics is not readily available (or perhaps is even impossible) is the famous Kochen–Specker theorem,[18] which is formulated and proved within standard mathematics, in particular in the environment of (standard) real and complex numbers. It is important to stress that the reasons are internal for mathematics, i.e. it is mathematics itself which imposes certain restrictions. So, it is difficult (or perhaps even impossible) to distinguish the features of the physical world from the features of mathematical formalism that are responsible for this non-realist interpretation of quantum mechanics.

In order to overcome these difficulties, some authors propose "transferring" the standard theories into a different mathematical environment, i.e. into topos theory. From their perspective, standard physics can be viewed as just a special case of a more general approach: it operates within Sets, which is just one of many possible topoi. A reformulation within a more general framework is meant to be "neo-realist" "in the sense that physical quantities are represented by arrows $A_\phi : \Sigma_\phi \to R_\phi$ and propositions are represented by sub-objects of Σ_ϕ, the set of which is a Heyting algebra. In this sense, these topos-based theories all 'look' like classical physics, except of course that, generally speaking, the topos concerned is not Sets" [3, p. 26]. Of course, several technical difficulties have to be overcome, but once TQT is available it offers a way of giving a (neo-)realistic interpretation of quantum mechanics. "This neo-realism is the conceptual fruit of the fact that, from a categorial perspective, a physical theory expressed in a topos 'looks' like classical physics expressed in the topos of sets" [2, p. 14].

Obviously, these theories are formulated on a very general abstract level. They are not concerned with explanations in the (standard) sense characterized in Part 1: they do not aim to explain concrete, down-to-earth physical (biological, medical) phenomena in terms of, for instance, variational principles, fixed-point theorems, properties of differential equations, limit theorems for stochastic processes, etc. In

[17] Flori gives a succinct characterization of realism in the context of physics which is very convenient for our purposes: "By a 'realist' theory we mean one in which the following conditions are satisfied: (i) propositions form a Boolean algebra and (ii) propositions can always be assessed to be either true or false. As will be delineated in the following, in the topos approach to quantum theory both of these conditions are relaxed, leading to what Isham and Döring called a neo-realist theory" [5, p. 3].

[18] "However, standard quantum theory precludes any such naive realist interpretation of the relation between formalism and the physical world. And this obstruction comes from mathematical formalism itself in the guise of the famous Kochen–Specker theorem which asserts the impossibility of assigning values to all physical quantities whilst, at the same time, preserving the functional relations between them" [2, p. 49]. For a presentation and discussion of the Kochen–Specker theorem, see e.g. [8].

particular, we would hardly expect new predictions to be made (at the empirically accessible level).[19] On the other hand, it cannot be denied that category-theoretic accounts impact our understanding of physical theories. So, what is their explanatory contribution? It is reasonable to claim that the (possible) explanatory contribution is located on the metatheoretical rather than the theoretical level.

Consider a toy example concerning number theory: for instance, investigations into the structure of prime numbers.[20] Do we really better understand prime numbers after we make findings concerning non-standard models for PA? Broadly speaking, there are two answers: (1) No, not in the least! We are interested in the relevant, mathematical phenomena, i.e. in the behavior of genuine natural numbers, not in artifacts of weak theories. So, we should rather look for "The Theory of Genuine Numbers" and not bother about its (impoverished) variants and their pathological models. (2) Yes, and it is crucial! In this way we learn something about the place of our understanding of natural numbers within the "space of possible conceptualizations". Metamathematical investigations concerning non-standard models for PA might also yield new purely mathematical results: there are examples of mathematical problems being solved with metamathematical (for instance, model-theoretical) methods.[21]

This is a simple example, but it exhibits the general problem: how does analysis of the "space of possible descriptions/theoretical accounts" enhance our understanding of the phenomena in question, i.e. on the object level? Metatheoretical reflection on the status and properties of the conceptual system constitutes the most important part of what we call "the philosophy of X".

Turning back to TQT, it might be claimed that it reveals the true reasons why a realistic interpretation of quantum mechanics is (at least) difficult, or perhaps even impossible (as the Kochen–Specker theorem is claimed to show): the reason is that standard formulations "live" in the very special topos of classical sets, which imposes limitations on the possibilities of interpreting the notion of physical proposition. In particular, the disturbing Kochen–Specker theorem can be loosely interpreted as an accidental feature of the standard topos (Sets). After getting rid of these limitations, a novel formulation (and interpretation) can be found. Such contributions have a clear metatheoretical character: they do not focus on concrete phenomena, but rather on the ways of conceptualizing the problem.[22]

---

[19] At least today. One day even very abstract results in CT might have such predictive power; however, this is just a speculative remark—it seems much more probable that CT will remain at the level of a "theory of theories".

[20] These facts might have bearing on the security of banking systems (so in this sense, number-theoretic problems have a physical interpretation) or at least some (indirect) impact on the real world.

[21] Similarly, we describe a mechanical phenomenon by a differential equation (and this is empirically adequate). We can also analyze this equation within a non-standard setting, e.g. within non-standard analysis. Would these investigations contribute to our understanding of the phenomenon, given there are no empirically detectable differences (e.g. different predictions)?

[22] One important philosophical issue is the realism/antirealism problem (also in quantum mechanics), but this does not affect the problem of applying the theory as a predictive tool (where empirical

Undeniably, there is an increase in understanding, but can it be accounted for in terms of the notion of explanation? Ultimately, the answer depends on the general attitude towards the role and epistemic input of metatheoretical investigations. The contributions of CT are to be found on the metatheoretical level, and even if they do not play any explanatory role in the standard sense, they might contribute to our philosophical or maybe metaphysical understanding both of the world and of our methodology of doing science. The level of generality is high, so it is perhaps more justified to consider these explanations as belonging to metaphysics (or philosophy of science), rather than to physics per se. But regardless of the answer, these investigations contribute to a better understanding of the very notion of realism and of the related question: what features of standard mathematics make the realistic approach to quantum mechanics difficult?

3.4 Conclusion

The answer to the main question depends on our understanding of the notion of mathematical explanation. I argued for the thesis that in the most standard sense encountered in the philosophical literature, CT offers no explanations of physical phenomena: it operates on a very high level of generality, and it offers no tools for describing (and, in particular, predicting) specific phenomena. In comparison with, for example, topological graph theory (which provides an important class of mathematical explanations), it really seems to be a kind of "abstract nonsense".

However, CT undoubtedly offers important tools for analyzing concepts and theories. In particular, TQT contributes to a better understanding of the question of which features of physical theories (couched in standard mathematical formalism) are a hindrance for the realistic interpretation of quantum mechanics. Being "a theory of theories" rather than an object-level theory, it contributes to our understanding of physics, albeit on a very profound and abstract level.[23] It could be fruitful to confront it with the idea of abstract explanations in science. Broadly speaking, these are explanations expressed in the form of abstract properties: "Abstract explanations involve an appeal to a more abstract entity than the state of affairs being explained. I show that the abstract entity need not be causally relevant to the explanandum for

accuracy is the criterion). So, these deep issues might not have any direct impact on practical matters, but they are nevertheless important for the understanding of the field.

[23] "Category theory is not a theory of everything. It is more like, as topologist Jack Morava put it, [...] "a theory of theories of anything". In other words, it is a model of models. It leaves each subject alone to solve its own problems, to sharpen and refine its toolset in the ways it sees fit. That is, CT does not micromanage in the affairs of any discipline. However, describing any discipline categorically tends to bring increased conceptual clarity, because conceptual clarity is CT's main concern, its domain of expertise. ...Finally, category theory allows one to compare different models, thus carrying knowledge from one domain to another, as long as one can construct the appropriate "analogy", i.e., functor" [25, p. 400].

its features to be explanatorily relevant" [21, p. 857].[24] Using this terminology, we might think of CT as providing "meta-abstract properties", i.e. higher-level abstract properties that concern not even abstract properties of the phenomena, but abstract properties of the conceptual systems and theories that describe the phenomena. So, CT offers "meta-abstract explanations". In particular the fact that quantum physics is not easily compatible with realist interpretation (or is perhaps even incompatible) is explained by the "meta-abstract property" of topos Sets. Whether this is in any sense also a property of the world is to be discussed, but surely it brings about some understanding of the conceptual systems we use.

It is somewhat paradoxical that the formalism which aims at a (neo-)realistic description of the quantum world uses non-classical logic.[25] This stands in contrast to the approaches to the realism/antirealism issue, which define this distinction in semantic terms (the bivalence of classical logic being a defining property of realism). Assume that it is not possible to remain within classical mathematics (and physics) and give a coherent realist interpretation of quantum mechanics. Does this claim express a claim about the physical world which is not otherwise expressible (see footnote 5), or is it just a claim concerning formalism? Or perhaps it is rather a contribution to our understanding of the notion of "realistic theory"? At the fundamental level, the borderline between questions that belong to physics proper or to methodology and philosophy is somehow blurred. The *explanandum* is not a particular physical phenomenon, but a methodological phenomenon: in the discussed case, it is the impossibility of giving a coherent realistic interpretation of QM. So perhaps the explanatory contribution of CT should be called philosophical or metaphysical, rather than physical. Its explanatory force is revealed on a very abstract level of analysis.

Acknowledgements The preparation of this paper was supported by an National Science Centre grant 2016/21/B/HS1/01955. I would like to thank Professor Zbigniew Semadeni for his very helpful comments.

References

1. M. Darrason, Mechanistic and topological explanations in medicine: the case of medical genetics and network medicine. Synthese **195**(1), 147–173 (2018)
2. A. Döring, C. Isham, "What is a thing?": topos theory in the foundations of physics, in *New Structures for Physics*, vol. 813, Springer Lecture Notes in Physics, ed. by B. Coecke (Springer, Heidelberg, 2008), pp. 753–937. arXiv:0803.0417

[24] Program explanations (conceived as a kind of abstract explanation) might also be mentioned in this context. A seminal paper is [11]; the authors claim that "not efficacious itself, the abstract property was such that its realization ensured that there was an efficacious property in the offing" [11, p. 114]. For the application of this account to mathematics, see [17]; for a discussion in the context of relativistic computation, see [27].

[25] However, it is not obvious that TQT achieves its own aim of being a realist theory; for a critical assessment, cf. [4].

3. A. Döring, C. Isham, A topos foundation for theories of physics: I. formal languages for physics. J. Math. Phys. **49**(5) (2008)
4. B. Eva, Topos theoretic quantum realism. Br. J. Philos. Sci. **68**(4), 1149–1181 (2017)
5. C. Flori, *A First Course in Topos Quantum Theory*, vol. 868, Springer Lecture Notes in Physics (Springer, Berlin, 2013)
6. T.C. Hales, The honeycomb conjecture. Discret. Comput. Geom. **25**(1), 1–22 (2001)
7. H. Halvorson (ed.), *Deep Beauty Understanding the Quantum World Through Mathematical Innovation* (Cambridge University Press, Cambridge, 2011)
8. Held C, The Kochen-Specker theorem, in *The Stanford Encyclopedia of Philosophy*, ed. by E.N. Zalta, spring 2018 edn (Metaphysics Research Lab, Stanford University, 2018)
9. H. Herrlich, H-E. Porst (eds.), *Category Theory at Work* (Heldermann Verlag, Berlin, 1991)
10. P. Huneman, Diversifying the picture of explanations in biological sciences: ways of combining topology with mechanisms. Synthese **195**(1), 115–146 (2018)
11. F. Jackson, P. Pettit, Program explanation: a general perspective. Analysis **50**(2), 107–117 (1990)
12. D. Kostić, Mechanistic and topological explanations: an introduction. Synthese **195**(1), 1–10 (2018)
13. M. Kuś, B. Skowron, K. Wójtowicz, Why Categories? (2019), in this volume
14. E. Landry, *Categories for the Working Philosopher* (Oxford University Press, Oxford, 2017)
15. M. Lange, What makes a scientific explanation distinctively mathematical? Br. J. Philos. Sci. **64**(3), 485–511 (2013)
16. D. Liggins, Abstract expressionism and the communication problem. Br. J. Philos. Sci. **65**(3), 599–620 (2013)
17. A. Lyon, Mathematical explanations of empirical facts, and mathematical realism. Australas. J. Philos. **90**(3), 559–578 (2012)
18. A. Lyon, M. Colyvan, The explanatory power of phase spaces. Philos. Math. **16**(2), 227–243 (2008)
19. P. Mancosu. Explanation in mathematics, in *The Stanford Encyclopedia of Philosophy*, ed. by E.N. Zalta, summer 2018 edn. (Metaphysics Research Lab, Stanford University, 2018)
20. J-P. Marquis. What is category theory? in *What is category theory?*, ed. by G. Sica (Polimetrica International Scientific Publisher, Monza/Italy, 2006), pp. 221–255
21. C. Pincock, Abstract explanations in science. Br. J. Philos. Sci. **66**(4), 857–882 (2015)
22. C. Rathkopf, Network representation and complex systems. Synthese **195**(1), 55–78 (2018)
23. T. Räz, On the application of the honeycomb conjecture to the bee's honeycomb. Philos. Math. **21**(3), 351–360 (2013)
24. T. Räz, The silent hexagon: explaining comb structures. Synthese **194**(5), 1703–1724 (2017)
25. D. Spivak. Categories as mathematical models, in *Categories for the Working Philosopher*, ed. by E. Landry (Oxford University Press, Oxford, 2017), pp. 381–401
26. D.J. Watts, S.H. Strogatz, Nature **393**(6684), 440 (1998)
27. K. Wójtowicz, The significance of relativistic computation for philosophy of mathematics. In *Hajnal Andréka and István Németi on Unity of Science: From Computing to Relativity Theory through Algebraic Logic* (to appear), Madarasz J., Szekely G. (eds.), Springer-Verlag (2019)
28. S. Yablo, Abstract objects: a case study. Philos. Issues **12**, 220–240 (2002)

Chapter 4
The Application of Category Theory to Epistemic and Poietic Processes

Józef Lubacz

Abstract The goal of this paper is to explore the potential for applying the notional framework of category theory to some aspects of human activity associated with acquiring knowledge (epistemic activity), as well as to activity of a sort resulting in the creation of artefacts of any kind (poietic activity). These kinds of activity are conceptualized as processes. The ideas and constitutive components of epistemic and poietic processes are introduced, and against this background possible applications of category theory to the analysis and monitoring of progress in the unfolding of such processes is considered.

4.1 Introduction

The terms *epistemic process* and *poietic process* are used to refer to two types of human activity: namely, that which aims at acquiring knowledge and that which results in artefacts. While this broad interpretation of the ancient Greek terms *episteme* and *poiesis* is not quite in accordance with their original meaning and application, we have nevertheless chosen to employ them here because we have found no better alternative—especially where the term *poiesis* is concerned (such that it would be on a par with the more well-established *episteme*).

Acquiring knowledge is something that pertains to both natural matters and artefacts of any kind (be they technological, economic, social, cultural, artistic, etc.). The activity resulting in specific artefacts requires a special kind of knowledge (i.e. technical knowledge), and many forms of knowledge (e.g., scientific knowledge) are themselves artefactual, so in principle, epistemic and poietic activities are not independent. We shall not attempt to explore the ancient Greek understanding of episteme as knowledge associated with achieving truth, as opposed to *phronesis* (practical knowledge), *techne* (arts-and-crafts-based knowledge), *nous* (intuitive knowledge) and *sophia* (wisdom); this is because viewed from the standpoint of

J. Lubacz (✉)
International Center for Formal Ontology, Warsaw University of Technology, Warszawa, Poland
e-mail: jl@tele.pw.edu.pl

M. Kuś and B. Skowron (eds.), *Category Theory in Physics, Mathematics, and Philosophy*, Springer Proceedings in Physics 235,
https://doi.org/10.1007/978-3-030-30896-4_4

our own sophisticated and diversified contemporary understanding of the relations between knowledge and truth, such an opposition does not seem tenable [4]. We shall also not seek to address the problem of distinguishing natural objects from artefacts, on account of the fact that finding general criteria of demarcation for this has proved problematic [6], and is anyway becoming of diminishing importance in modern societies populated not just by "pure artefacts", but also by hybrids of the latter and natural objects.

Epistemic and poietic activities will be conceptualized here as processes. We do not seek to formalize the notion of process, but we will distinguish the *pattern* according to which a process evolves from the *actualization* of that pattern resulting in consecutive states of the process. We do not attempt to determine whether the process pertains to an individual subjective activity, or to some collective intersubjective activity, and we also do not refer to any "paradigms" or "programs" of activity. What we shall be focusing on is the constitutive elements of the patterns of epistemic and poietic processes, and relations between the two types of process. In other words, our aim is to consider general, generic properties of such processes.

We believe that exploring the relevance of category theory (CT) to epistemic and poietic processes can furnish a somewhat different perspective on the role and applications of the theory to that which emerges when it is considered in relation to mathematics, logic, physics or computer science. In particular, this is a consequence of the emphasis placed on the interrelation of epistemic and poietic processes—which, according to our best knowledge, is seldom noticed or analysed. Where reflection on knowledge acquisition is concerned, most of the currently predominant tendencies concentrate on such traditional epistemological issues as the theoretical versus the empirical, realism versus antirealism, perception versus observables, observables versus unobservables, theories versus models, etc. Moreover, these longstanding problems may themselves be considered from the perspective of epistemic and poietic processes.

4.2 Epistemic and Poietic Processes

Philosophical thought concerning observed processes of change pertaining to various aspects of the world and its inhabitants has a long tradition; in Western philosophy its origin is usually associated with Heraclitus. Process philosophy, especially as a consequence of its revival in the last century, has become a complex and highly diversified field of reflection [8]. Because of this diversification, there seems to be no agreed-upon, paradigmatic conceptualization of the notion of process. Process philosophy has developed partly in opposition to the longstanding and historically dominant tradition of philosophical reflection focused on essences of things. Both essentialist philosophy and process philosophy centre on ontological and metaphysical questions whose applicability to the issues essentially involved where epistemic and poietic processes are concerned is limited. More inspiration and ideas can be drawn from epistemology, but only rather indirectly, because epistemic issues have seldom had as their focus the processual and methodological aspects of acquiring knowledge—apart from those well-known aspects considered by the likes of Popper, Kuhn, Lakatos and Feyerabend, of course. In the case of poietic processes, one of the main sources of inspiration and ideas is, not surprisingly, to be found in engineering and design methodologies [5].

In most of the reflections of process philosophy, some process is invoked to describe and interpret changes in regard to some states of affairs pertaining to some entities or phenomena. Such changes, if observable, must somehow be distinguished. We shall refer to such distinguishable characteristics of changes as *states of a process*. The succession of states may be considered as appearing in continuous or discrete temporal moments.[1] Should we need to distinguish some special moments of time, such as are associated with certain special conditions internal or external to the process (associated, for example, with predefined states of the process or predefined time-limits in achieving some states), then the latter may be interpreted as embedded in the former.

The above is, in effect, a conceptualization of the *dynamicity* of a process. We assume that the dynamic features of a process—i.e. the transition from one state to another—takes place according to some *pattern* which is regarded as a static constituent of a process. The relation between the statics and the dynamics of processes may be interpreted by analogy with some classical philosophical concepts: for example, with Aristotle's potency and actualization, or Heidegger's analysis of being and Being. If the pattern of the process is interpreted as the essence of that process, then the relation between the dynamic and the static features of processes may be considered a nexus between essentialism and processism. These issues, however, lie beyond the scope of this paper.

In general, the pattern of a process may be time-dependent, and thus exhibit some form of dynamicity. In the following, we shall not ignore this, but will assume that the pattern does not change within time-intervals long enough to enable meaningful analysis of the interplay of the pattern and its actualization that results in dynamic features of the process. The notion of a *process pattern* may be variously interpreted. If a process is used to conceptualize, for instance, the evolution of nature, then the existence of a pattern may be doubted in circumstances where the dynamicity of the changes is considered to come from a pure interaction of some sub-processes which themselves exhibit no predefined patterns. At the same time, this position is only tenable if it refers to a pattern defined a priori. However, one cannot reasonably claim that processes are pattern-free on an a posteriori basis, as that would mean that the process state transitions are "completely random": i.e. that no regularities whatsoever can be detected. The problem with this is that "complete randomness" is a theoretical notion, and so is unrealistic when applied to the conceptualization of real-world processes (such as the evolution of species or the phenomena of quantum mechanics).

[1] In general, states of a process need not appear in time: a succession of such states can be indexed in an abstract way—i.e. in a time-independent fashion. We shall, however, not enter further into discussion of this issue.

With the above clarification in place, we may now proceed to introduce the proposed conceptual structure of the pattern and dynamicity of epistemic and poietic processes. We shall not aim at achieving a formalized definition, but rather concentrate on the intuitive side. First of all, it is necessary to clarify how a static process pattern (*PP*) results in a given succession of state transformations—with the latter being further referred to as a *process-actualization* (*PA*). If no divine intervention is assumed, then a *PP* causes a *PA* as an effect of some direct or indirect (e.g. machine-aided) human intervention, this intervention being called the animation of a process (*AP*). The *PA* results from the *PP* as an effect of the *AP*. Animation will be the nexus between the static and dynamic features of epistemic and poietic processes.[2]

Let us first consider poietic processes. The actualization of a process (*PA*) may be divided into two consecutive time-related phases (borrowing standard terms from engineering): the design phase and the implementation phase. In the design phase what is to be implemented (i.e. brought into existence or reality, realized) is described in some way; in the implementation phase the description is transformed into the object of description. The implementation transforms, so to say, "words into flesh", which is a quite miraculous activity, and one which even St. John did not explain. (Well, he didn't have to!)[3] We likewise shall not go into this intriguing issue, and so shall concentrate on the design phase.

We propose to conceptualize the design phase of a poietic process as follows:

- OP—*description* of the object of poiesis, i.e. of what is to be implemented.
- PO—*description* of how the implemented object is perceived (directly, or indirectly with some observational and/or experimental aids).
- QC—*description* of the evaluative and qualitative criteria of the poietic process, pertaining to OP and PO, and to the required relation between OP and PO.
- No particular form of the descriptions is assumed, as the form may, and in reality it does, depend on the type of the object of poiesis and on the adopted methodology of design.
- The OP plays the role of a prescription for the object of poiesis.
- PO serves as the reference point with respect to which the implemented object is evaluated using QC criteria (i.e. the evaluation is indirect).
- The state-transformation of the process is interpreted as the change of OP, PO and QC: individually, pair-wise, or all three together.
- When the design process is initiated, the PO is empty (contains no information) until the designed object is implemented. The transformations of OP in this time-interval (e.g., adding, in effect, new aspects and details to the description) are evaluated with respect to criteria whose role is to monitor the required features of OP in accordance with some predefined methodological paradigm; such criteria form a subset of QC, further referred to as *design*-QC (d-QC). The d-QC criteria are also active when the PO is not empty.

[2]We should mention that in some applications, especially in information science and engineering, patterns are considered types subject to instantiation rather than animation.

[3]Paraphrasing the "word into flesh" metaphor, the epistemic process may be said to transform "flesh into word".

- The remaining criteria of the QC pertain to the required relation of OP and PO, which for a poietic process establishes some predefined interpretation of how well the implemented object satisfies the OP—i.e. the prescription. The latter criteria will be referred to as *poietic-satisfaction*-QC (s-QC).
- The PO is constructed during the evolution of the process alongside the activity of evaluating the object being implemented (as the intermediate and/or final result of the implementation). In some domains of application of poietic processes (and also epistemic processes), the activity of evaluation is referred to in terms of testing (active or passive), verification and validation.

In the above conceptualization, it is assumed that the poietic process results in some entity that is external to the mind: i.e. that the object of poiesis is not a mental product. This is assumed also with respect to the objects of epistemic processes considered below. In fact this assumption can be dropped, resulting only in some quite non-essential alterations to the conceptual framework introduced. We will not, however, consider such a scenario, as it requires the introduction of some phenomenological and quasi-phenomenological notions pertaining to forms of consciousness [4]—something which goes beyond the scope of what is directly relevant to the main issues of this paper.

We shall use the conceptual framework introduced for poietic processes, after some additional interpretation and modifications, for epistemic processes, too (i.e. processes of acquiring knowledge). The notion of knowledge is a complex notion which may be understood in various ways—in a narrower and a broader sense, from various viewpoints, and with various aspects being taken into account [4]. Here, we shall restrict our considerations to those forms of knowledge contemporarily associated with science and technology: i.e. forms of knowledge thought reliable in the sense that they are subject to objectified criteria of "quality". Obviously, this kind of knowledge may be treated as acquired through some form of poietic process: i.e. a processes of constructing artefacts.

Given the above, an epistemic process can be conceptualized as shown below. Here, the abbreviations used earlier for components of poietic processes will be appended with a letter "e" (OPe substitutes for OP, POe substitutes for PO, etc.).

- The product of an epistemic process is a description OPe of an object of some episteme—i.e. of something that acquiring knowledge aims at. In contrast to the OP description of the poietic process, which performs the role of a prescription, OPe may be said to play the role of a particular way of expressing some *hypothesis* or other.
- OPe is evaluated in terms of the description POe of how the object of the episteme is perceived (directly, or indirectly with some observational and/or experimental aids).
- The evaluation employs qualitative criteria QCe, pertaining to OPe and POe, as well as to the required relation between OPe and POe.
- POe, like PO, is constructed during the course of the evolution of the epistemic process, together with the evaluation of OPe in terms of POe—i.e. the evaluation of the hypothesis as expressed in OPe.

- The evaluation refers to some predefined criteria of *epistemic-satisfaction* s-QCe, for how well the OPe hypothesis corresponds to what is expressed in POe.
- Note that OPe is evaluated in terms of POe, in contrast to the case of poietic processes, in which PO is evaluated in terms of OP.[4]
- As with the role of the design d-QC criteria for poietic processes, the d-QCe criteria are used to monitor the required features of the evolution of the epistemic process in accordance with some predefined methodological paradigm.

Epistemic processes have thus been conceptualized as a special, modified case of poietic processes; yet we could have approached this the other way around, defining poietic processes as derivative of epistemic processes. Whatever the case, however, such a procedure emphasizes the close conceptual relationship between the two. Note that this also comes to light when we consider that the construction of PO and POe, and the evaluation of such processes, involves activity of both an epistemic and a poietic character. The two types of processes are thus interdependent; nevertheless, they can and should be distinguished for theoretical reasons and for practical purposes.

4.3 The Potential of Category Theory

With the preparatory considerations explored in the preceding sections now in place, we are in a position to consider the application of category theory (CT) to epistemic and poietic processes. Over the course of the several decades of its development, CT has been applied not only to purely mathematical issues, but also to logic, physics, computer science and philosophy, as evidenced by the publication of this volume. This demonstrates, so to say, the conceptual power and potential of CT, which seem to have been foreseen by the founding fathers of the theory in their seminal paper [1]: emphasis was put there on the problem of how different kinds of generally understood mathematical structure could be mapped one to another, the structures were named "categories", and the name chosen for the mappings was "functors".[5] We wish to stress the generality of the original conception of CT, as it is precisely this generality that is the source of the usefulness of CT to different areas of research, including those dealing with epistemic and poietic processes themselves.

Above all, however, it must be clearly stated that the apparatus of CT can only be employed for those conceptual components of epistemic and poietic processes which are expressible in some formal language and form—based, possibly indirectly, on some formalism of mathematics and/or logic. Obviously, this is not the case for all areas of science and technology (not to mention, for instance, artistic activity) in which epistemic and poietic processes function. In the area of science, physics is least problematic in this respect; on the other hand, it may be difficult to use CT to

[4]This resembles, respectively, the *mind-world* and *world-mind* "directions of fit" discussed by Searle [7].

[5]The term "functor" was apparently employed earlier by Carnap, while the use of the term "category" was probably inspired by Aristotle or Kant.

conceptualize the epistemic processes associated with some important aspects of, say, biology and economics. This pertains also to those poietic processes that are such as to render the formulation of design objectives and features with OP, PO and QC in a formalized way is difficult or even unrealistic.[6] In most real-life design, and the implementation associated with most artefacts, such features are expressed in a mixture of formal and natural language. Bearing this in mind, in order to make CT practically applicable, the non-formalized constituents of OP, PO and QC should be clearly distinguished—at least for those kinds and aspects of artefacts that can be formalized. Another issue is that some diverse aspects of artefacts may require a different formalization language and form: e.g., physical and functional aspects, which cannot be unified in one common, uniform formalism [3]. In that case, the CT formalism will have to be applied separately to each aspect.

Keeping in mind the above constraints, we shall now consider those fundamental concepts of CT that, together with its basic formalism, can apparently be adapted—with appropriate interpretation, modification, or extension—to epistemic and poietic processes.

During the transition of a poietic process from state to state, the prescription OP of the designed entity[7] undergoes consecutive transformations: e.g., $OP_1 \rightarrow_1 OP_2 \rightarrow_2 OP_3 \rightarrow_3 \ldots$ The question is this: do the consecutive versions of OP describe the same entity? This is a special case of the question of identity, which has a long tradition in philosophy. If all the instances of OPi are expressed as categories, and the arrows are functors (morphisms), then the answer to the question could be as follows: what is preserved in the sequence are some structural features determined by the OPi and the kind of mapping between them determined by the $\rightarrow_i$. The answer thus pertains only to the structural features of the entities designed, so the next question could be this: is the preservation of structural features enough to give a positive answer to the identity question? Well, the answer may depend on the kind of entity being designed. Consider, for example, entities that are common in information and communication technology: i.e. objects that are a composition of software and hardware. The software component determines the functionality of the entity, while the hardware is the component which enables animation of the functionality. Many aspects of functionality may be expressed in terms of structural relations between specific morphisms of a category. Thus, the identity question may potentially be positively answered with respect to functionality with the aid of a CT conceptualization. The hardware component is another issue—touched on in passing below. Note only that in many cases the objective of the design is to determine functionality in a way that is independent of implementation, in order to allow for different hardware realizations of the functionality [2].

The identity issue pertains to epistemic processes in a similar way, although not in quite the same sense. Here a sequence of descriptions $OPe_1 \rightarrow e_1\ OPe_2 \rightarrow$

[6]Note, however, that even in the case of works of art some features can formalized: e.g. rules of perspective in paintings.

[7]In the following we use the expression "designed entity" instead of "object of design", in order to avoid confusion with objects of categories.

$e_2\, OPe_3 \rightarrow e_3 \ldots$ may be interpreted as pertaining to some development of a hypothesis, expressed in some particular way or other (in the form of a theory, set of sentences, model, etc.), that is about the entities to which the process pertains. CT may be helpful in determining the criteria of identity appropriate for some aspect or phenomenon associated with the entity being considered. We can also ask whether, or on what conditions, the sequence converges; this is in essence a version of one of the classical problems of epistemology. Note that such a problem does not arise in poietic processes. Indeed, there is also another classical problem of epistemology that does not pertain to such processes: the problem of competing theories. In the latter case, with the help of a CT conceptualization, one can consider criteria that would allow one to judge whether the competing theories pertain to the same entity, or are compatible in some sense; this is a variant of the general problem of identity.

Resolving the problems considered above would require one to construct some super-structures pertaining to functors—categories of functors, of a kind appropriate for epistemic and poietic processes. It is important to take account of the fact that the above identity problems are naturally associated with evaluating and monitoring the progress of a process towards desired ends: i.e. with testing, verification and validation. The methodological aspects of the progress of the sequence $OP_1 \rightarrow_1 OP_2 \rightarrow_2 OP_3 \rightarrow_3 \ldots$ is monitored with respect to the d-QC criteria in the case of a poietic process, and with respect to the d-QCe criteria in the case of the $OPe_1 \rightarrow e_1\, OPe_2 \rightarrow e_2\, OPe_3 \rightarrow e_3 \ldots$ sequence of an epistemic process.

The consecutive elements of the $OPe_1 \rightarrow e_1\, OPe_2 \rightarrow e_2\, OPe_3 \rightarrow e_3 \ldots$ sequence are also evaluated with respect to the POe description with the s-QCe criteria, in order to ensure that the process meets the desired epistemic goal. In the case of a poietic process it is the PO that is evaluated with respect to the consecutive elements of the $OP_1 \rightarrow_1 OP_2 \rightarrow_2 OP_3 \rightarrow_3 \ldots$ sequence with the s-QC criteria, in order to ensure that the process results in the implantation of the designed artefact. The application of the CT conceptualization raises problems essentially analogous to those described above.

Note that so far we have not mentioned objects of categories. The fact that objects can be eliminated from the definition of a category has interesting implications. With respect to poietic processes this allows for the possibility of maintaining a state of ignorance concerning the inner structure of objects—something which, as was mentioned above, is instrumental in separating the functional features of the designed entities from the enablers of the functions (e.g., separation of software from hardware in the design). Note that from this perspective, the classical essentialism versus structuralism controversy loses much of its supposed importance and relevance.

In the case of epistemic processes, the irrelevance of the inner structure of objects of categories has other interesting consequences, which are related to the problem of observables versus unobservables. In principle, the POe is a direct or indirect description of observables, while the OPe description may refer to observables, but also to unobservables (where the status of the latter may be only temporary). It seems that CT conceptualization would be well suited to expressing such circumstances with the aid of functors: the objects of the OPe category that are assumed to be unobservable can be mapped onto observable objects of the POe category. This kind of mapping is especially important when it comes to evaluating an OPe description with respect to a POe description with the s-QCe criteria.

We have not sought to suggest any specific features of categories that would be suitable when seen from the perspective of the problems presented above. This is partly because much surely depends on the particular type and area of application of the epistemic and poietic processes in question, but also because we believe that CT conceptualization will prove flexible enough to be adapted to a broad range of particularities. An example of this flexibility is the application of special kinds of category—namely, topoi—to the investigation of the new and intriguing concepts involved in quantizing gravity. From the perspective of the problems touched on in this paper, it will be especially interesting to consider topoi for which, in principle, the logical law of excluded middle is substituted with some alternative of, say, an intuitionistic sort, in order to take account of real-life testing procedures that yield such results as "pass", "fail" or "cannot decide".

A final comment: every category may be interpreted in terms of a graph (objects = nodes, morphisms = oriented edges), with functors treated as mappings between such graphs. This is a useful circumstance because—as has been well documented—graph-based presentations tend to enhance our capacity for comprehending even the most complex conceptual constructs.

References

1. S. Eilenberg, S. MacLane, General theory of natural equivalences. Trans. Am. Math. Soc. **58**(2), 231–294 (1945)
2. W. Houkes, P.E. Vermaas, *Technical Functions: On the Use and Design of Artefacts*, vol. 1 (Springer Science & Business Media, Berlin, 2010)
3. P. Kroes, Engineering and the dual nature of technical artefacts. Camb. J. Econ. **34**(1), 51–62 (2009)
4. Z. Król, J. Lubacz, The subject's forms of knowledge and the question of being, in *Contemporary Polish Ontology*, ed. by B. Skowron (De Gruyter, Berlin, 2019)
5. G. Parsons, *The Philosophy of Design*, vol. 1 (Polity Press, Cambridge, 2016)
6. B. Preston, Artifact, in *The Stanford Encyclopedia of Philosophy*, 2018th edn., ed. by E.N. Zalta (Metaphysics Research Lab, Stanford University, 2018)
7. J.R. Searle, *Intentionality: An Essay in the Philosophy of Mind* (Cambridge University Press, Cambridge, 1983)
8. J. Seibt, Process philosophy, in *The Stanford Encyclopedia of Philosophy*, 2018th edn., ed. by E.N. Zalta (Metaphysics Research Lab, Stanford University, 2018)

Chapter 5
Asymmetry of Cantorian Mathematics from a Categorial Standpoint: Is It Related to the Direction of Time?

Zbigniew Semadeni

Abstract Category theory is symmetric in the sense that all definitions, theorems and proofs have uniquely defined duals, obtained by formal reversing of arrows. In contrast, the products and coproducts in typical categories whose objects are sets endowed with basic algebraic, topological etc. structures of *Cantorian Mathematics* show a specific lack of symmetry. A philosophical question is raised: What features of mathematics and mathematical thinking are related to this phenomenon? Some hints suggest that this is related to the role of the concept of a function in mathematics and the domination of many-to-one thinking. This in turn may be attributed to implicit thinking in terms "causes precede the effects" and to the arrow of time.

5.1 Introduction

Initially (since 1945) category theory was regarded as a convenient conceptual language for certain aspects of mathematical theories. Later, however, new ideas developed by Lawvere and others showed that topos theory could provide a unified framework for set theory, logic and a good part of mathematics [13]. Consequently, category theory was viewed as a new contender for a foundation of mathematics along with set theory, or—as Mac Lane would put it—as a proposal for the organization of Mathematics [17, pp. 398–407], [18, p. 331].

The purpose of this paper[1] is to use categorial concepts to highlight a certain feature of *Cantorian Mathematics*; this term refers here to basic mathematical structures of algebra, topology, functional analysis etc. expressed in terms of set theory, as they were conceived prior to the emergence of category theory, i.e., by the middle

[1]The present paper is based on a talk delivered at the conference "Category Theory in Physics, Mathematics and Philosophy" held at the Warsaw University of Technology, 16–17 November 2017. Some ideas presented here were published in [23].

Z. Semadeni (✉)
Institute of Mathematics, University of Warsaw, ul. Banacha 2, 02-097 Warsaw, Poland
e-mail: semadeni@mimuw.edu.pl

M. Kuś and B. Skowron (eds.), *Category Theory in Physics, Mathematics, and Philosophy*, Springer Proceedings in Physics 235,
https://doi.org/10.1007/978-3-030-30896-4_5

of the 20th century. We will consider categories in which objects are sets[2] provided with specific structures, while morphisms are structure preserving maps (homomorphisms, continuous maps etc.).

General category theory is fully symmetric in the sense that each definition, each theorem and each proof has its uniquely defined **dual**, obtained by the process: *reverse all arrows*, that is, by the following replacements in the theory:

- each expression of the type $\alpha\beta = \gamma$ is replaced by $\beta\alpha = \gamma$;
- in each morphism the word "domain" is replaced by "codomain" and "codomain" is replaced by "domain", that is, each $\alpha\colon A \to B$ is replaced by $\alpha\colon B \to A$;
- arrows and composites are reversed, while the logical terms are unchanged [15], [16, pp. 31–33], [24].

Each statement of the theory has a unique dual statement, e.g., the dual of "α is monic" (i.e., is a monomorphism) is "α is epic", and vice versa; the dual of "A is an initial object" is "A is a terminal object"; the dual of a product is a coproduct.

In an axiom system for category theory, the dual of each axiom is also an axiom. Consequently, in any proof of a theorem, replacing each statement by its dual gives a valid proof of *the dual theorem*. This is the *duality principle* in category theory.

We will show that—from this point of view—Cantorian mathematics is specifically asymmetric.

5.2 Products and Coproducts

As a crucial example we consider the notion of a product of an indexed family of objects $\{A_t\}_{t\in T}$ in a category $\mathcal{E}$, defined as an object P together with a family $\{\pi_t\colon P \to A_t\}_{t\in T}$ of morphisms (called projections) having the unique factorization property: for every object X and every family of morphisms $\{\xi_t\colon X \to A_t\}_{t\in T}$ there exists a unique morphism $\theta\colon X \to P$ such that the diagrams commute, i.e., $\pi_t\theta = \xi_t$ for $t \in T$. If a product exists, it is unique up to commuting isomorphism.

A coproduct (also called a categorial sum) of a family of objects $\{A_t\}_{t\in T}$ is defined dually as an object S together with a family $\{\sigma_t\colon A_t \to S\}_{t\in T}$ of morphisms (called injections) such that for every object X and every family of morphisms $\{\xi_t\colon A_t \to X\}_{t\in T}$ there exists a unique morphism $\theta\colon S \to X$ such that $\theta\sigma_t = \xi_t$ for $t \in T$.

5.2.1 Products in Certain Categories of Sets with Structures

Most of the following examples are concrete categories, i.e., categories $\mathcal{E}$ equipped with a faithful functor $U : \mathcal{E} \to$ **Set**. Symbols of specific categories used here gen-

[2] It is assumed here that all sets, functions, topological spaces etc. considered here are small sets, [16, p. 22].

erally follow those of [16, p. 12]. In the category **Set** of (small) sets and functions, the product of $\{A_t\}_{t\in T}$ is the cartesian product $A = \prod_{t\in T} A_t$ with the coordinate projections $\pi_t\colon A \to A_t$ $(t \in T)$.

In the category **Top** of topological spaces and continuous maps, in its full subcategory **Comp** of compact Hausdorff spaces, in the category $\mathbf{Set}_*$ of sets with selected base-points and base-point preserving functions, in the analogous category $\mathbf{Top}_*$, in the category **Ord** of partially ordered sets and non-decreasing maps, in the categories **Grp**, **Ab** and **AbComp** of groups (resp. abelian groups and compact abelian groups) and their homomorphisms (resp. continuous homomorphisms)—in all these categories the product $\prod_{t\in T} A_t$ is the cartesian product endowed with a suitable structure (in **Ord** it is the cardinal product in the sense of [2, I. Sect. 7], [24, 3.3.8]).

The category $\mathbf{Ban}_1$ of Banach spaces and linear contractions, i.e., linear operators of norm $\|T\| \le 1$ (called also short linear operators), may appear to be an exception to the rule, as the cartesian product of infinitely many Banach spaces is not a Banach space. In fact, their product is the ℓ^∞-product consisting of all $\{x_t\}_{t\in T}$, $x_t \in X_t$, such that $\sup_{t\in T} \|x_t\| < \infty$. However, this exception may be regarded as spurious. If we adjust the concept of the carrier of the Banach space structure, replacing the whole vector space X by its closed unit ball $\{x \in X\colon \|x\| \le 1\}$ (which actually determines the geometric structure of the whole space) and replacing the category $\mathbf{Ban}_1$ by the category $\mathbf{Ban}_{\bigcirc}$ of closed unit balls and restrictions of linear contractions, then the product object becomes simply the cartesian product of balls.

Another exception is the category **AbTor** of abelian torsion groups. The direct product $A = \prod_{t\in T} A_t$ of such groups need not be a torsion group, e.g., $\mathbb{Z}_2 \times \mathbb{Z}_3 \times \mathbb{Z}_5 \times \ldots$ However, the torsion-subgroup P of A consisting of all torsion elements (i.e., all elements of finite order) is a product in **AbTor**, [1, 10.20].

5.2.2 *Coproducts in the Same Categories*

In contrast to the preceding, coproducts may be markedly different from each other. This is clearly shown in the following Table 5.1.

In the categories **Set**, **Top**, $\mathbf{Top}_*$, **Comp** and $\mathbf{Set}_*$ coproducts are based on the same construction, namely on the disjoint union $A = \bigsqcup_{t\in T} A_t$, defined as $\bigcup_{t\in T}(A_t \times \{t\})$, with obvious injections $\sigma_t\colon A_t \to A$. In **Top** the coproduct $\bigsqcup_{t\in T} A_t$ is equipped with the disjoint union topology [6, Chap. 2, Sects. 2, 4]; in $\mathbf{Top}_*$ the coproduct is the wedge sum, i.e., the quotient of the disjoint union obtained by identifying the base points to a single point; in **Comp** copoducts are the Stone–Čech compactifications of disjoint unions. In **Ord** it is the cardinal sum [2, I. Sect. 7], [24, 3.3.10].

However, in other categories coproducts may differ basically. In **Grp** the coproduct is the free product of groups, whereas in its full subcategory **Ab** the coproduct is the direct sum $\bigoplus A_t$ (called also the external direct sum); thus the **Ab**-coproduct of two copies of the cyclic group $\mathbb{Z}$ is commutative, whereas their **Grp**-coproduct is not commutative (moreover, its center is trivial).

Table 5.1 In the left-hand column there are symbols of categories together with concise descriptions of their objects and morphisms. In the middle column, for each category there is a description of products. Conspicuously, the cartesian product $\prod$ appears in each cell with exception of Banach spaces and commutative C^*-algebras (however this difference disappears when one changes the definition of the carrier of the object taking the closed unit ball instead of the whole vector space). In the right-hand column there are descriptions of respective coproducts. There are six distinct types of them: coproducts related to the disjoint union $\bigsqcup A_t$; free products $\coprod A_t$ of groups; coproducts related to directs sums $\bigoplus A_t$ of abelian groups; coproducts related to tensor products $\bigotimes A_t$ of rings; ℓ^1-sums; a specific construction for automata

Symbol of the category objects, morphisms	**Product of a family** $\{A_t\}_{t\in T}$ **of objects**	**Coproduct of a family** $\{A_t\}_{t\in T}$ of objects
Set sets functions	Cartesian product $\prod_{t\in T} A_t$	Disjoint union $\bigsqcup_{t\in T} A_t = \bigcup_{t\in T} (A_t \times \{t\})$
Top topological spaces continuous maps	Cartesian product $\prod A_t$ with product topology	Disjoint union $\bigsqcup A_t$ $A_t \times \{t\}$ open-and-closed
Top$_\bullet$ pointed topological spaces based maps	Cartesian product $\prod A_t$ with base point $\{\bullet_t\}_{t\in T}$	Wedge sum $\bigvee A_t = \left(\bigsqcup A_t\right)/\sim$ with quotient topology
Comp compact spaces continuous maps	Cartesian product $\prod A_t$ with product topology	Stone-Čech compactification of the disjoint union $\beta\left(\bigsqcup A_t\right)$
Gr groups homomorphisms	Cartesian product $\prod A_t$ multiplication componentwise	Free product $\coprod A_t$ group of words
Ab abelian groups homomorphisms	Cartesian product $\prod A_t$ addition componentwise	(external) direct sum $\bigoplus A_t$ $x_t = e_t$ for almost all t
Abcomp compact abelian groups continuous homomorphisms	Cartesian product $\prod A_t$ product topology addition componentwise	Bohr compactification of the direct sum $\bigoplus A_t$ with coproduct topology
CRng commutative rings (with units) unit-preserving homomorphisms	Cartesian product $\prod R_t$ addition and multiplication componentwise	Tensor product $\bigotimes R_t$ of rings
Ban$_1$ Banach spaces linear operators $\|\|T\|\| \le 1$	ℓ^∞-product the set of all $\{x_t\}_{t\in T}$ satisfying $\sup_{t\in T} \|\|x_t\|\| < \infty$	ℓ^1-sum the set of all $\{x_t\}_{t\in T}$ satisfying $\sum_{t\in T} \|\|x_t\|\| < \infty$
C*algcom$_1$ commutative C^*-algebras unit-preserving homom	ℓ^∞-product the set of all $\{x_t\}_{t\in T}$ satisfying $\sup_{t\in T} \|\|x_t\|\| < \infty$	Injective tensor product $\widehat{\widehat{\bigotimes}} A_t$
Aut Mealy automata $\langle X, S, Y, \delta, \lambda\rangle$ where $\delta : S \times X \to S$ $\lambda : S \times X \to Y$	$\prod A_t, \prod S_t, \prod Y_t$ induced map δ induced map λ	A construction in terms of disjoint sums of finite cartesian products

In **AbComp** the coproduct is the Bohr compactification [12], [9, p. 430], [10], [22, pp. 249–254] of the direct sum $\bigoplus A_t$, provided with a coproduct topology [19]. This is particularly interesting in view of the Pontryagin duality. A locally compact abelian group G is compact if and only if its dual group $\widehat{G}$ (the group of characters, i.e., continuous homomorphisms $G \to \mathbb{R}/\mathbb{Z}$) is discrete [14, Chap. VII], [9, Sect. 24]; consequently, the category **Ab** (which may be regarded as that of discrete abelian groups) is equivalent (in the sense of [16, IV.4]) to the opposite category (i.e., dual) of **AbComp**. Thus, one might expect a somehow "dual behavior" of their products and coproducts; yet, the products in both categories are akin to Cartesian products and coproducts to direct sums.

In the category **C*algcom**$_1$ of commutative C^*-algebras with units and their homomorphisms, the product is an ℓ^∞-product while the coproduct is the injective tensor product $\widehat{\widehat{\bigotimes}} A_t$ (called also the weak tensor product), [22, pp. 355–361]. By the Gelfand duality theorem, **C*algcom**$_1$ is equivalent to the dual of **Comp** [22, Sects. 10.2, 12.6, 13.3], so the situation is analogous to that with products and coproducts in **Ab** and **AbComp**.

In the category **CRng** of commutative rings with units and unit-preserving ring homomorphisms the product of a family $\{R_t\}_{t\in T}$ is—as in any category of algebras of the same type—the Cartesian product $A = \prod A_t$ with suitable operations and coordinate projections, whereas the coproduct of this family is the tensor product of rings $\bigotimes R_t$ (i.e., the tensor product over $\mathbb{Z}$ for rings as $\mathbb{Z}$-algebras, [21, p. 65]).

In the category **Ban**$_1$ of Banach spaces and linear contractions the coproduct of a family $\{X_t\}_{t\in T}$ is its ℓ^1-sum consisting of all $\{x_t\}_{t\in T}$ such that $\sum_{t\in T} \|x_t\| < \infty$.

In the category **Aut** of finite Mealy automata $\langle X, S, Y, \delta, \lambda\rangle$ (where S denotes a set of states, X is an input alphabet, Y is an output alphabet, $\delta : S \times X \to S$ is a transition function, $\lambda : S \times X \to Y$ is an output function and morphisms are triples $\xi\colon X_1 \to X_2$, $\sigma\colon S_1 \to S_2$, $\eta\colon Y_1 \to Y_2$ such that suitable diagrams commute), the product of $\langle X_t, S_t, Y_t, \delta_t, \lambda_t\rangle$ is the triple $\prod X_t$, $\prod X_t$, $\prod X_t$ with the induced maps δ, λ and is related to that in **Set**, [5]. However, coproducts are sophisticated and quite different [25], [24, 2.4.7, 3.3.15].

5.2.3 *Recapitulation of Main Points*

The forgetful functors from each of the categories considered above to **Set** (or to **Set** $\times$ **Set** $\times$ **Set** in case of automata) commute with products and do not commute with coproducts.

A similar asymmetry, albeit in a much milder form, concerns equalizers and coequalizers [24, Sect. 3.5].

A consequence of the above asymmetry product–coproduct is an analogous asymmetry of limits (called also inverse limits or projective limits) and colimits (direct limits or inductive limits) of diagrams [16, pp. 62–72].

One may distinguish two kinds of categorial duality. One, which may be labeled as *syntactic*, mentioned above, is based on the formal replacing of each morphism $\alpha\beta$ (in a category $\mathcal{E}$) by $\beta\alpha$. The other, which may be labeled as *functional* (and is an anti-equivalence, i.e., equivalence with the opposite category $\mathcal{E}^{op}$), is based on constructions related to some contravariant hom-functor $\hom_{\mathcal{E}}(-, E_0)$. The first kind is significant in the general theory, whereas the second yields better insight into specific categories, like those discussed here.

5.3 A Philosophical Discussion

At this point a philosophical question arises: *What features of Cantorian Mathematics lie behind this asymmetry?*

Clearly, the membership relation: element–set, $x \in X$ is a basic asymmetry. However, this explanation is not adequate here, as the following examples show.

Let **Rel** denote the category of sets and binary relations. Objects are sets, a morphism $R\colon A \to B$ is a triple (R, A, B) where $R \subseteq A \times B$. If $S \subseteq B \times C$ is another such relation, the composite morphism is $(S \circ R, A, C)$, where

$$S \circ R = \{(a, c) \in A \times C \mid \exists_{b \in B} (a, b) \in R \text{ and } (b, c) \in S\}.$$

The empty set is the zero object. The coproduct of a family $\{A_t\}_{t \in T}$ of objects in **Rel** is the disjoint union $A = \bigsqcup_{t \in T} A_t$ with obvious injections $\sigma_t\colon A_t \to A$. The product is the same disjoint union $A = \bigsqcup A_t$ with morphisms $\pi_t\colon A \to A_t$ defined as the inverse relations $\pi_t = \sigma_t^{-1}$ for $t \in T$. The categorial symmetry of **Rel** suggests itself.

If a set P is partially ordered by a relation $\leq$ and is regarded as a category in which the morphisms $a \to b$ are exactly those pairs (a, b) for which $a \leq b$ and if the greatest lower bound $\inf\{a_t\}_{t \in T}$ of a family exists, then it is the product of that family. Analogously, $\sup\{a_t\}_{t \in T}$ is the coproduct. Here again the symmetry is clear.

The last two examples suggest that **the product-coproduct asymmetry** of the categories shown in the table above follows from **the asymmetry of many-to-one relationship in the notion of a function** $f\colon X \to Y$.

Now the next question arises as to why such **many→one** thinking predominates in Mathematics. Certainly it is deeply rooted in our minds. It is so in early arithmetic as, e.g., in $4 + 3 = 7$ the natural direction is from numbers 4, 3 to the sum 7. The opposite relation—decomposing a number into summands—is also important but definitely secondary. Most computations lead from given data to a result. Solving an equation appears to be a way backwards. In calculus, functions play a vital role, whereas their multivalued inverse relations are used only occasionally.

In the real life causes precede the effects [20, Chap. 7]. This is implicit in common thinking, manifests in ordinary language, and also shapes mathematical thinking. It is subordinated to the psychological arrow of time which—according to Hawking [8, Chap. 9]—is determined by the thermodynamic arrow of time. It is likely an evolutionary effect in mathematical thought.

In a preliminary search, in the context of discovery, the mathematician's thinking may have no preconceived direction, but systematic reasoning (as in a proof) has a clear direction (the case of backward reasoning, from the consequent to the antecedent, is usually an intentional, conscious reversing of the direction).

The concept of a function has two aspects: dynamic and static. The first, related to change and motion, was implicit in Newton's approach, continued till the 19th century, and still somehow influences the thinking in terms of functional dependence. Also an "input/output machine" approach, with permissible inputs and the corresponding outputs, is dynamic.

The static conception of a function developed slowly from Euler's analytic form to Dedekind's modern, purely logical and completely general notion of a many-to-one mapping from a set to a set [3, Chap. V–VII], [7, pp. 228–232].

By its very nature, set theory is static. In the set-theoretical approach, the dynamic conception of a function is replaced by a static relation, conceived as a set of pairs. Time, which played the role of a distinct variable in the 18th and 19th century, became one of the space coordinates in $\mathbb{R}^n$ [17, pp. 123–133]. Moreover, the New Math movements of the 1960s contributed to the attitude that time belongs to physics. On the other hand, even if formally Mathematics expressed in terms of set theory appears static, Cantorian models of physical processes represent a dynamic world.

The author is indebted to Prof. Jiří Rosický for paying attention to some important aspects of the question considered in the paper. Most of the examples discussed here are of an "algebraic" nature, i.e., they are given by operations whose general form is $TA \to A$ where $T : \mathbf{Set} \to \mathbf{Set}$ is a functor. But there are also structures of "coalgebraic" nature given by $A \to TA$, [4]. In such a case, coproducts are preserved but not products. A typical example are transition systems given by $A \to \mathcal{P}A$, where $\mathcal{P}$ is the power-set functor [11, part on non-deterministic automata]. Thus, the asymmetry considered in this paper applies to the "algebraic" part of Cantorian Mathematics while an opposite asymmetry applies to the "coalgebraic" part of mathematics. The former is predominant in classical mathematics (particularly in applications to physics) while coalgebraic part is mostly stimulated by Computer Science. The opposite one-to-many relation, for example decomposing a number into summands or multi-valued inverse of a function, is considered as being important but secondary in Cantorian Mathematics. While many-to-one is typical to it, one-to-many is typical to coalgebraic mathematical theories. The example of transition systems shows that one-to-many is not always given by some many-to-one. Here, it reflects the non-deterministic nature of a process where one has more ways how to go from a state of the system to another state.

The general theory of coalgebras requires category theory. Before the appearance of the latter, mathematicians could deal only with asymmetric Cantorian Mathematics; the coalgebraic part was hidden.

References

1. J. Adáimek, H. Herrlich, G. Strecker, *Abstract and Concrete Categories* (Wiley, New York, 1990). http://katmat.math.uni-bremen.de/acc/acc.pdf
2. G. Birkhoff, *Lattice Theory*. American Mathematical Society, vol. 25, 2nd ed. (Colloquium Publications, New York, 1948)

3. C.B. Boyer, *The History of the Calculus and its Conceptual Development* (Dover Publications, New York, 1959)
4. *Coalgebra*. https://en.wikipedia.org/wiki/Coalgebra
5. H. Ehrig, M. Pfender, *Kategorien und Automaten* (Walter de Gruyter, Berlin, 1972)
6. R. Engelking, *General Topology, Monografie Matematyczne*, vol. 60 (PWN-Polish Scientific Publishers, Warszawa, 1977)
7. J. Ferreirós, *Labyrinth of Thought. A History of Set Theory and its Role in Modern Mathematics* (Birkhäuser Verlag, Basel, 1999)
8. S.W. Hawking, *A Brief History of Time: from the Big Bang to Black Holes* (Bantam Books, Toronto, 1988)
9. E. Hewitt, K.A. Ross, *Abstract Harmonic Analysis*, vol. I (Springer, Berlin-Göttingen-Heidelberg, 1963)
10. P.J. Higgins, Coproducts of topological Abelian groups. J. Algebra **44**(1), 152–159 (1977)
11. B. Jacobs, *Introduction to Coalgebra: Towards Mathematics of States and Observation*. Cambridge Tracts in Theoretical Computer Science (Cambridge University Press, Cambridge, 2016). http://www.cs.ru.nl/B.Jacobs/CLG/JacobsCoalgebraIntro.pdf
12. S. Kaplan, Extensions of the Pontrjagin duality I: infinite products. Duke Math. J. **15**, 649–658 (1948)
13. F.W. Lawvere (ed.), *Toposes, Algebraic Geometry and Logic*. Lecture Notes in Mathematics, vol. 274 (Springer, Berlin–Heidelberg–New York, 1972)
14. L.H. Loomis, *An Introduction to Abstract Harmonic Analysis* (Van Nostrand, Toronto-New York-London, 1953)
15. S. Mac Lane, Duality for groups. Bull. Am. Math. Soc. **56**, 485–516 (1950). https://projecteuclid.org/download/pdf_1/euclid.bams/1183515045
16. S. Mac Lane, *Categories for the Working Mathematician* (Springer-Verlag, New York, 1971)
17. S. Mac Lane, *Mathematics. Form and Function* (Springer Verlag, New York-Berlin-Heidelberg, 1986)
18. S. Mac Lane, I. Moerdijk, *Sheaves in Geometry and Logic. A First Introduction to Topos Theory* (Universitext, Springer, 1992)
19. P. Nickolas, Coproducts of abelian topological groups. Topol. Appl. **120**(3), 403–426 (2002)
20. R. Penrose, *The Emperor's New Mind* (Oxford University Press, Oxford, 1989)
21. H. Schubert, *Categories* (Springer, Berlin-Heidelberg-New York, 1972)
22. Z. Semadeni, *Banach Spaces of Continuous Functions*. vol. I. Monografie Matematyczne. vol. 55 (PWN-Polish Scientific Publishers, Warszawa, 1971)
23. Z. Semadeni, Is the Cantorian mathematics symmetric?, in *Theory of Sets and Topology. In Honour of Felix Hausdorff (1868–1942)* ed. by G. Asser et al. (VEB Deutscher Verlag Wissenschaften, Berlin, 1972), pp. 467–471
24. Z. Semadeni, A. Wiweger, *Einführung in die Theorie der Kategorien und Funktoren*, Teubner-Texte zur Mathematik. BSB B. G. Teubner Verlagsgesellschaft, Leipzig (1979). http://d-nb.info/800278038 [translated from the Polish second edition]
25. A. Wiweger, On coproducts of automata. Bull. Acad. Polonaise Sci. **21**, 753–758 (1973)

Chapter 6
Extending List's Levels

Neil Dewar, Samuel C. Fletcher and Laurenz Hudetz

Abstract Christian List (Noûs, forthcoming, 2018, [24]) has recently proposed a category-theoretic model of a system of levels, applying it to various pertinent metaphysical questions. We modify and extend this framework to correct some minor defects and better adapt it to application in philosophy of science. This includes a richer use of category theoretic ideas and some illustrations using social choice theory.

6.1 List's Descriptive, Explanatory, and Ontological Levels

In general, a *system of levels* for List [24] is a preordered class: that is, a class $\mathscr{L}$ equipped with a reflexive and transitive binary relation $\leq$. The elements of $\mathscr{L}$ are to be interpreted as *levels*, and the binary relation $\leq$ as the relation of *supervenience*: $L \leq L'$ means L' supervenes on L. Thus, requiring that $\leq$ be a preorder amounts to assuming that every level supervenes upon itself, and if the level L_1 supervenes on the level L_2, and L_2 on the level L_3, then L_1 supervenes on L_3. In requiring only these characteristics, List is deliberately opting for a fairly weak conception of supervenience. For instance, it is possible to have two distinct levels within the system, neither of which supervene upon the other, or to have two distinct levels both of which supervene upon the other.

This framework gives us the resources to consider certain relationships between systems of levels. Given two systems of levels $\mathscr{L}$ and $\mathscr{L}'$, a function $f : \mathscr{L} \to \mathscr{L}'$

N. Dewar
LMU Munich, Geschwister-Scholl-Platz 1, 80539 Munich, Germany
e-mail: neil.dewar@lrz.uni-muenchen.de

S. C. Fletcher (✉)
University of Minnesota, Twin Cities, 271 19th Ave S, Minneapolis, MN 55455, USA
e-mail: scfletch@umn.edu

L. Hudetz
London School of Economics, Houghton Street, London WC2A 2AE, UK
e-mail: l.hudetz@lse.ac.uk

M. Kuś and B. Skowron (eds.), *Category Theory in Physics, Mathematics, and Philosophy*, Springer Proceedings in Physics 235,
https://doi.org/10.1007/978-3-030-30896-4_6

is a *monotonic* map if it preserves the order relation: i.e., for any levels $L_1, L_2 \in \mathscr{L}$, if $L_1 \leq L_2$ then $f(L_1) \leq f(L_2)$. If there are monotonic maps $f : \mathscr{L} \to \mathscr{L}'$ and $f' : \mathscr{L}' \to \mathscr{L}$ that are mutually inverse to one another, then $\mathscr{L}$ and $\mathscr{L}'$ are said to be *isomorphic*, or structurally equivalent. If $\mathscr{L} \subseteq \mathscr{L}'$ and there is some map $f : \mathscr{L} \to \mathscr{L}'$ such that $L_1 \leq L_2$ if and only if $f(L_1) \leq f(L_2)$, then $\mathscr{L}$ is a *subsystem* of $\mathscr{L}'$.

Such is List's general framework. There is one significant difference between our presentation of this framework so far and his: we have not yet mentioned *categories* at all. Although we will start to use category-theoretic language below, we have avoided it so far to make clear that the abstract framework in his [24] can be understood without category theory.

Let us now turn to more specific kinds of systems of levels. First, consider the case of a system of *ontological* levels. In such a system, each level is associated (or identified) with a set of *possible worlds* for that level. If one level supervenes upon another, then there is a (unique) *supervenience map* from the subvenient level to the supervenient level; this map is required to be surjective.[1] We also impose the following two requirements:

1. for any level L, the identity map is the supervenience map from L to itself; and
2. if σ is the supervenience map from L_1 to L_2, and σ' is the supervenience map from L_2 to L_3, then $\sigma' \circ \sigma$ is the supervenience map from L_1 to L_3.

More compactly stated, a system of ontological levels forms a *concrete posetal category* in which every function is surjective. To say that it is a *concrete category* means that it is a class of sets, equipped with functions between those sets that are closed under composition and include all identity functions. To say that it is a *posetal* category means that between any two sets, there is at most one function. Although this compact description uses category-theoretic apparatus, it only does so in the form of appeal to concrete categories (i.e., categories of sets and functions). This means that the category-theoretic language provides a convenient way to express our requirements, rather than being an indispensable tool; if desired, we could do everything purely in set-theoretic terms, as our initial description of a system of levels in terms of a preordered class evinces.[2]

Two specific kinds of systems of ontological levels are worth mentioning. The first is what List calls a *system of levels of grain*. For this, we take as given a set Ω of "possible worlds". Each level in the system is given by some partition of Ω: one such partition supervenes upon another just in case every cell of the former is a union of cells of the latter, with the supervenience map taking each "fine-grained" cell to the "coarse-grained" cell of which it is a part. Since any such map is guaranteed to be surjective, it follows that a system of levels of grain is an ontological system of levels.

[1] In Sect. 6.3 we will consider whether this requirement is justified.

[2] Of course, there is a sense in which this is true for any application of category-theoretic apparatus, at least insofar as one can represent a category as a set-theoretic structure. But this trivial sense is not the one we have in mind here.

List [24, p. 10] claims that even though "a system of ontological levels is formally more general than a system of levels of grain, there exists a functor from any system of ontological levels to some system of levels of grain." In case the system of ontological levels has a lowest level—a level on which all others supervene—the desired functor maps that level bijectively onto a class representing the lowest-level possible worlds, and all other levels are mapped to partitions of these worlds induced by the supervenience maps of the system of ontological levels. In case there is no lowest level, List proposes to construct one formally using an inverse limit of the ontological levels. "For a posetal category," he writes, "an inverse limit can always be constructed, though we need not interpret it as anything more than a mathematical construct."

This last statement must be qualified, however. Inverse limits are defined only for posetal categories each pair of whose elements has a greatest lower bound [1, p. 194],[3] so an inverse limit can be constructed in a system of ontological levels when any two levels have a common level on which they both supervene that itself supervenes on all such common subvenient levels. A simple example of a system of levels in which this does not occur consists of two levels, L_1 and L_2, that only supervene on themselves. No inverse limit can be constructed for this system, and it does not mirror any system of levels of grain.

Perhaps there is some other construction that allows one to exhibit a functor from a large class of—if not *any*—system of ontological levels to some system of levels of grain. In the previous simple example, for instance, it is quite natural to define a system of levels that adds a new level to L_1 and L_2, whose worlds are the Cartesian product of the worlds of each of those two levels, and a pair of supervenience maps for the two components' projection maps.[4] This could then be generalized to any system of ontological levels that has a set-sized number of "lowest levels," and might be given a category theoretic expression using enriched categories. But as systems of ontological levels are currently defined, they includes systems with a (proper) class of lowest levels, whose worlds cannot be so combined into a set-sized Cartesian product.

The second specific kind of system of ontological levels that List considers is a *system of descriptive levels*—roughly, a system of ontological levels in which each level is the set of worlds describable in a specific language. More precisely, a language $\mathbf{L}$ is defined as a set of elements—formal expressions called *sentences*—that are equipped with a negation operator $\neg : \mathbf{L} \to \mathbf{L}$ and a bifurcation of the power set $\mathscr{P}(\mathbf{L})$ into two, labeled "consistent" and "inconsistent." It is required in particular that:

1. any set containing a sentence and its negation is inconsistent;

[3] In more detail, an inverse limit in a category $\mathfrak{C}$ can be characterized as the limit of a functor from a partially ordered set, considered as a small category, to $\mathfrak{C}$, and such limits exist when $\mathfrak{C}$ has small products and equalizers [9, Theorem 2.8.1]. Posetal categories trivially always have equalizers, but their products are just greatest lower bounds.

[4] Thanks to Christian List for this suggestion. Note that the Cartesian product is the product in the category of sets, not in a posetal category—see footnote 3.

2. inconsistency is preserved by taking supersets;
3. $\varnothing$ is consistent; and
4. every consistent set is contained in a maximal consistent set (i.e., a consistent set containing, for every sentence $\phi \in \mathbf{L}$, either ϕ or $\neg\phi$).

The *ontology* for $\mathbf{L}$, denoted $\Omega_{\mathbf{L}}$, is defined as the set of all maximal consistent subsets of $\mathbf{L}$. Each such subset (i.e., each element of $\Omega_{\mathbf{L}}$) is a *world*. A sentence $\phi \in \mathbf{L}$ is said to be *true at* a world $w \in \Omega_{\mathbf{L}}$ if $\phi \in w$, and the *propositional content* of ϕ, denoted $[[\phi]]$, is defined as the set of all worlds at which ϕ is true.

A system of descriptive levels is thus a system of ontological levels, each of which is the ontology of some language. Within such a system, a higher-level sentence $\phi' \in \mathbf{L}'$ is defined (by List) to be *reducible* to a lower-level sentence $\phi \in \mathbf{L}$ if and only if the propositional content of ϕ is the inverse image of ϕ' under σ, the supervenience map from $\Omega_{\mathbf{L}}$ to $\Omega_{\mathbf{L}'}$. And he defines the higher level of description L' to be reducible to the lower level of description L if every sentence of the higher level's associated language $\mathbf{L}'$ is reducible to some sentence of the lower level's associated language $\mathbf{L}$.

Note that, so defined, not every system of descriptive levels will be one in which the higher levels reduce to the lower levels: as List observes, this provides a sense in which supervenience does not entail reduction. For a concrete example [13], let $\mathbf{L}'$ be the propositional language for level L' whose only sentence-letter is F, and let $\mathbf{L}$ be the propositional language for level L with sentence-letters $\{P_0, P_1, \dots\}$, each equipped with the standard notion of consistency. $\Omega_{\mathbf{L}'}$ only contains two worlds: ω_F, which contains F, and $\omega_{\neg F}$, which contains $\neg F$. Let $\omega \in \Omega_{\mathbf{L}}$ be the world containing every P_i. Define $\sigma : \Omega_{\mathbf{L}} \to \Omega_{\mathbf{L}'}$ as follows: for any $\widetilde{\omega} \in \Omega_{\mathbf{L}}$,

$$\sigma(\widetilde{\omega}) := \begin{cases} \omega_F & \text{if } \widetilde{\omega} = \omega, \\ \omega_{\neg F} & \text{otherwise.} \end{cases}$$

Then the levels $\Omega_{\mathbf{L}}$ and $\Omega_{\mathbf{L}'}$ equipped with the maps σ, $\mathrm{Id}_{\Omega_{\mathbf{L}}}$, and $\mathrm{Id}_{\Omega_{\mathbf{L}'}}$ constitute a system of descriptive levels.

Now observe that $\sigma^{-1}([[F]]) = \{\omega\}$. But $\{\omega\}$ is not a definable subset of $\Omega_{\mathbf{L}}$ since there is no sentence $\phi \in \mathbf{L}$ such that $[[\phi]] = \{\omega\}$.[5] So the sentence $F \in \mathbf{L}'$ is not reducible to any sentence in $\mathbf{L}$; thus, L' is not reducible to L. We will discuss in Sect. 6.2, however, what sort of conditions we could place on supervenience that would associate it with reduction.

[5] *Proof* Suppose for *reductio* that $[[\phi]] = \{\omega\}$. Since ϕ is a finite sentence, not every sentence letter can occur in it. So suppose P_i does not occur in ϕ. Then since $\omega \models \phi$, it must be the case that $\omega' \models \phi$, where ω' is just like ω save that $P_i \notin \omega'$ (as the truth value of a sentence in propositional logic is dependent only on the truth values of the sentence letters occurring in it). But then $\omega' \in [[\phi]]$, although $\omega' \neq \omega$, so we have a contradiction. □

6.2 Supervenience and Reduction

As discussed in Sect. 6.1, for List, supervenience need not entail reduction, in the sense that there can be systems of levels of description in which the levels are not reducible to one another. In this section, we look at how, by imposing certain assumptions on the system of descriptive levels, we can recover this entailment.

Before we begin, one preliminary observation is in order. For List, the worlds in a system of levels of description are identified as maximally consistent sets of sentences of a formal language, where the notion of a "language" is left very abstract. In this section's analysis, we will assume that the languages we are working with are first-order languages; this will play an important role when we invoke Beth's theorem. However, the languages may be many-sorted. We will also identify the worlds with models of the language, rather than maximal consistent sets of sentences—note that maximal consistent sets of sentences correspond to equivalence classes of elementarily equivalent models. This assumption is more for convenience than anything else; we do not believe that anything of great significance hangs on it.

Now, consider two levels of description L and L' associated with languages $\mathbf{L}$ and $\mathbf{L}'$, respectively. Let σ be a supervenience map from L to L'. We assume two conditions on the relationship between these levels. First, we assume that they are *compatible* with respect to σ, in the sense that if the vocabularies of $\mathbf{L}$ and $\mathbf{L}'$ intersect, then for any $\omega \in \Omega_{\mathbf{L}}$, $\omega|_{\mathbf{L}\cap\mathbf{L}'} = \sigma(\omega)|_{\mathbf{L}\cap\mathbf{L}'}$. This has the consequence that for any $\omega \in \Omega_{\mathbf{L}}$, we can define a unique expansion to a structure of signature $\mathbf{L} \cup \mathbf{L}'$: let $\omega + \sigma(\omega)$ be the structure such that for any symbol S in the vocabulary of $\mathbf{L} \cup \mathbf{L}'$ (including sort symbols),

$$S^{\omega+\sigma(\omega)} = \begin{cases} S^{\omega} & \text{if } S \in \mathbf{L}, \\ S^{\sigma(\omega)} & \text{if } S \in \mathbf{L}'. \end{cases}$$

Let $\Omega_{\mathbf{L}} + \Omega_{\mathbf{L}'} := \{\omega + \sigma(\omega) : \omega \in \Omega_{\mathbf{L}}\}$. This can be viewed as the union of the two levels relative to the supervenience map σ. Note that when its antecedent is satisfied, this assumption entails that ω and $\sigma(\omega)$ share a domain: or, in other words, that the lower-level theory already asserts the existence of higher-level objects (but without explicitly saying what they are like). Such an assumption is reasonably plausible if (for instance) we take the lower-level theory to include some kind of mereological theory that asserts the existence of mereological sums or fusions. Alternatively, one could seek to weaken this assumption by invoking methods for defining new sorts[6]; although such a project would be interesting, we do not undertake it here.

Our second assumption is that this class is *characterizable*, in the sense that there is a set of sentences T such that $\Omega_{\mathbf{L}} + \Omega_{\mathbf{L}'} = \mathrm{Mod}(T)$. This means that the union of the two levels comprises worlds that can be characterized as those in which certain

[6]See, in particular, [2, 6].

sentences in the union of their languages are true. This assumption is reasonably strong: in the context of psychology and physics, for example, it amounts to the assertion that there is some joint psycho-physical theory such that a distribution of psychological and physical facts is possible if and only if it is in accord with the joint theory. However, it is still nontrivially distinct from directly assuming reducibility. One way to obtain such a joint theory would be to conjoin our psychological and physical theories with a set of bridge laws connecting them; but the assumption of characterizability does not presume that the joint theory takes this "pre-reduced" form. Note further that, in general, assuming characterizability will mean that the antecedent of the first assumption is satisfied: that is, that there is at least one sort of object that both levels describe (and so, per the first assumption, about which they agree). This is plausible for realistic cases, in which we expect the two levels of description to share at least some vocabulary (e.g., empirical or observational terms).

Given these assumptions, it follows that the vocabulary of the higher level $\mathbf{L}'$ is *implicitly defined* by T in terms of the lower-level vocabulary $\mathbf{L}$: that is,

Proposition 1 *For any models* $\widetilde{\omega}_1$ *and* $\widetilde{\omega}_2$ *of* T, *if* $\widetilde{\omega}_1|_{\mathbf{L}} = \widetilde{\omega}_2|_{\mathbf{L}}$ *then* $\widetilde{\omega}_1 = \widetilde{\omega}_2$.

Proof Suppose that $\widetilde{\omega}_1|_{\mathbf{L}} = \widetilde{\omega}_2|_{\mathbf{L}}$. By the assumption of characterizability, there are $\omega_1, \omega_2 \in \Omega_{\mathbf{L}}$ such that $\widetilde{\omega}_1 = \omega_1 + \sigma(\omega_1)$ and $\widetilde{\omega}_2 = \omega_2 + \sigma(\omega_2)$. It follows that $\widetilde{\omega}_i|_{\mathbf{L}} = \omega_i$ and $\widetilde{\omega}_i|_{\mathbf{L}'} = \sigma(\omega_i)$ (for $i = 1, 2$). Hence,

$$\widetilde{\omega}_2|_{\mathbf{L}'} = \sigma(\omega_2) = \sigma(\widetilde{\omega}_2|_{\mathbf{L}}) = \sigma(\widetilde{\omega}_1|_{\mathbf{L}}) = \sigma(\omega_1) = \widetilde{\omega}_1|_{\mathbf{L}'}.$$

Thus, $\widetilde{\omega}_1 = \widetilde{\omega}_2$. □

Next, by Beth's theorem, it follows that L' is *explicitly defined* by T in terms of $\mathbf{L}$.[7] In the case of single-sorted languages, this means that for any (n-place) relation symbol R of $\mathbf{L}'$, there is an $\mathbf{L}$-formula τ_R such that $T \vdash \forall x_1 \ldots \forall x_n(Rx_1 \ldots x_n \leftrightarrow \tau_R(x_1, \ldots, x_n))$ and similarly for other kinds of symbols. The case of many-sorted languages is a bit more complex.[8] But it follows in either case that for every $\mathbf{L}'$-sentence ϕ', there is an $\mathbf{L}$-sentence ϕ such that $T \vdash (\phi' \leftrightarrow \phi)$. So ϕ' reduces to ϕ. Hence, it follows that L' reduces to L.

Thus, although supervenience maps between different levels (in general) are not associated with reduction, any supervenience map which fulfills our assumptions of compatibility and characterizability will be so associated.

An alternative way of establishing a relationship between supervenience and reduction proceeds not by imposing constraints on the supervenience map between levels, but rather by treating supervenience as a feature of sets of properties rather than worlds. Note that this is more in line with how supervenience is often defined in the literature: for example, as the Stanford Encyclopedia of Philosophy's entry on Supervenience [26] begins, "A set of properties A supervenes upon another set B

[7]For discussion of Beth's theorem, see [17].

[8]For a generalisation of Beth's theorem to many-sorted logics, see [2].

just in case no two things can differ with respect to A-properties without also differing with respect to their B-properties." In other words, the A-properties supervene upon the B-properties if any two possible worlds which have the same distribution of B-properties also have the same distribution of A-properties.

To make things more precise, let us focus on the notion of *strong global supervenience* [26, Sect. 4.3.2]: the A-properties strongly globally supervene upon the B-properties if and only if for any worlds w_1 and w_2, any B-preserving isomorphism between w_1 and w_2 is an A-preserving isomorphism between them.[9] Let us suppose that the A-properties are those expressed by a higher-level language $\mathbf{L}'$, and the B-properties are those expressed by a lower-level language $\mathbf{L}$. And suppose that our worlds (structures of signature $\mathbf{L} \cup \mathbf{L}'$) are characterizable, where this means the same thing as before: there is a theory T such that the worlds are the models of the theory. It then follows that

Proposition 2 *The A-properties strongly globally supervene on the B-properties if and only if T implicitly defines $\mathbf{L}'$ in terms of $\mathbf{L}$.*

Proof From left to right, suppose that the A-properties supervene upon the B-properties: that is, that for any worlds (models of T) ω_1 and ω_2, if f is an isomorphism from $\omega_1|_\mathbf{L}$ to $\omega_2|_\mathbf{L}$, then f is an isomorphism from ω_1 to ω_2. Now suppose further that we have two models ω_1 and ω_2 such that $\omega_1|_\mathbf{L} = \omega_2|_\mathbf{L}$. Clearly, then, the identity is an isomorphism between $\omega_1|_\mathbf{L}$ $\omega_2|_\mathbf{L}$, hence must be an isomorphism between ω_1 and ω_2; it follows that $\omega_1 = \omega_2$. So T implicitly defines $\mathbf{L}'$ in terms of $\mathbf{L}$.

From right to left, suppose that T implicitly defines $\mathbf{L}'$ in terms of $\mathbf{L}$, and let ω_1 and ω_2 be models of T such that f is an isomorphism from $\omega_1|_\mathbf{L}$ to $\omega_2|_\mathbf{L}$. For *reductio*, suppose that f is not an isomorphism from ω_1 to ω_2, so for some $R \in \mathbf{L}'$, $f[R^{\omega_1}] \neq R^{\omega_2}$. But now define ω_1' as follows: it has the same domain as ω_2, and for every $P \in \mathbf{L} \cup \mathbf{L}'$, $P^{\omega_1'} = f[P^{\omega_1}]$. By construction, f is an isomorphism from ω_1 to ω_1'.[10] So ω_1' is a model of T. And since f is an isomorphism from $\omega_1|_\mathbf{L}$ to $\omega_2|_\mathbf{L}$, it follows that $\omega_1'|_\mathbf{L} = \omega_2|_\mathbf{L}$, and so that $\omega_1' = \omega_2$. But then it follows that $R^{\omega_1'} = f[R^{\omega_1}] = R^{\omega_2}$, so we have obtained a contradiction. Hence, f must be an isomorphism from ω_1 to ω_2. □

From here, the analysis goes as before (i.e. via Beth's theorem): hence, if the higher-level properties supervene upon the lower-level properties, then (at least for one natural way of formalizing what this means) it will be accompanied by a reduction from the higher level to the lower level.

---

[9] As McLaughlin and Bennett [26] discuss, there are other notions of global supervenience one can define. For a compelling case that strong global supervenience is the most appropriate precisification of the intuitive notion of global supervenience, see Shagrir [32].

[10] This is an instance of what Button and Walsh [11] call the "push-through" construction.

6.3 Extensions: Partial and Non-surjective Maps

The first extension to List's framework we consider will be to *partial* supervenience maps. To motivate it, consider the case of a system of ontological levels as described in Sect. 6.1. These levels consist of possible worlds, each of which provides a full specification of the facts particular to that level. List takes the maps between levels, representing supervenience relations, to be *functions*, explaining that

> Supervenience means that each lower-level world determines a corresponding higher-level world: the lower-level facts, say the physical ones, determine the higher-level facts, say the chemical ones. By fixing all physical properties, we necessarily fix all chemical properties, in this example [24, p. 7].

However, the glosses before and after the colon are not equivalent: that the chemical facts, say, may supervene on the physical facts just means that there cannot be any difference between two chemical worlds without a difference between the physical worlds in the chemical worlds' supervenience bases—the preimages of the supervenience map on the two chemical worlds. But this is entirely compatible with there being some physical worlds that do not determine *any* non-trivial chemical worlds—worlds for which the empty set is not a logical model of any of their descriptions. Indeed, we expect there not to be any non-trivial chemical facts at all determined by those of a roiling quantum vacuum, or any non-trivial biological facts at all determined by the chemical facts of the atmosphere of Venus. Insofar as each of these is respectively a way a physical and chemical world could be, there can be physical worlds on which no non-trivial chemical world supervenes, chemical worlds on which no non-trivial biological world supervenes, etc.

List [24, p. 9] already cites approvingly a similar remark by Kim[11] regarding the difficulties of entity-based (rather than world-based) conceptions of levels, so this small extension seems welcome. Indeed, one can easily integrate into the definition of a system of ontological levels its consequence that the supervenience maps need only be partial: their domains of definition needn't be *all* the worlds at a level. (This doesn't by itself conflict with any of the formal properties defining a system of levels as a posetal category, but one must decide exactly how to model a category with partial morphisms: see, e.g., Cocketta and and Lackb [12, Sect. 1] for references to a number of options.) Since a system of levels of description is just a system of ontological levels with added structure, the same conclusion applies for them. The procedure for embedding, via a functor, a system of ontological levels into a system of levels of grain also remains the same; the image under the functor of any partial supervenience map will have some equivalence class of worlds in its lower-level domain that will just not have any image in the higher-level codomain within the system of levels of grain.

One way to regain totality for the supervenience maps would be to introduce a *null world* at each level, a world devoid of non-trivial facts. Then one could extend each

[11] Namely, that "not every 'complex' of 'lower-grade' entities will be a higher entity; there is no useful sense in which a slab of marble is a higher entity than the smaller marble parts that make it up" [20, p. 11].

previously partial supervenience function by mapping all the elements previously outside its domain of definition to the null world in its codomain. For consistency, the image of any null world under a supervenience map would then have to be a null world. This is coherent with the interpretation of the maps: a world devoid of non-trivial physical facts determines a world devoid of non-trivial chemical facts, and there can be no difference in the latter without a difference in the former—i.e., without the introduction of some non-trivial physical facts or other. However, such null worlds should be interpreted cautiously: if one wants the worlds at a level to be models of a theory describing that level, the null world will not be included if the theory has any existentially quantified axioms. So it probably should receive a fixed interpretation as a technical convenience rather than as a genuine possible world.

This second extension is to *non-surjective* supervenience maps. Now, List claims surjectivity just follows from the meaning of supervenience[12]:

> To say that the chemical level supervenes on the physical, or that the biological supervenes on the chemical, is to say that the class $\mathcal{S}$ of supervenience mappings contains one such mapping, $\sigma : \Omega \rightarrow \Omega'$, from the relevant lower level to the high one, where σ maps *each* lower-level world $\omega \in \Omega$ to the higher-level world $\omega' \in \Omega'$. We then call ω a *lower-level realizer* of ω'. The surjectivity of σ means that there are no possible worlds at the higher level that lack a possible lower-level realizer [24, p. 8].

So it does; but why should the *formal apparatus* of a system of levels necessarily commit to the metaphysical thesis that the *whole* higher level does so supervene? Even though we take the supervenience of levels to be plausible, we do not think that *characterizing* systems of levels require assuming it. Indeed, dropping this assumption, one finds the *interpretation* of the levels, the worlds comprising them, and the supervenience maps between them to be hardly different.

The key insight for this is that a supervenience map relates collections of facts at a lower level to facts at a higher level, hence need not relate all worlds at either level at all. If a supervenience map is not surjective, then there will be at least some higher level world that does not supervene on any lower level world in the map's domain. But still the map characterizes exactly which higher-level properties (and worlds) supervene on lower-level ones. That it is not surjective is just to indicate precisely which of these higher-level properties do not so supervene. The map represents a higher *level* supervening on a lower *level* if and only if it is surjective. This is important for applications in philosophy of science: When the worlds of levels are models of empirical theories, the failure of surjectivity indicates that the possibilities that a higher-level theory describes outstrip those that one committed to level-supervenience would expect. This can motivate revising the higher-level theory by restricting what it allows as possible, or generalizing the lower-level theory to allow more possibilities that could provide new supervenience bases for higher-level possibilities.

Furthermore, examples of non-surjective supervenience maps can be found in systems of different geometric levels, each of which is able to describe a wide range

[12]For emphasis we have italicized the word "each" in the passage.

of geometric possibilities.[13] Let us focus on the metric and the topological levels. There is a robust sense in which topological properties (such as being an open set or being a continuous function) supervene on metric properties (distances between points). Every metric naturally induces a topology and isometric metric spaces induce homeomorphic topological spaces. So in the jargon of possible worlds, sameness of worlds on the subvening metric level entails sameness of worlds on the supervening topological level. However, the supervenience map from the metric to the topological level (which sends each metric space to the topological space induced by it) is not surjective because not all topological spaces are metrizable. The topological level of description is essentially more general than the metric level, but nonetheless there is a supervenience map between them. Dropping the constraint of surjectivity enables us to capture such cases of supervenience. Thus, the framework gains expressive power without losing any advantages.

One further consequence of dropping surjectivity for the supervenience maps concerns List's argument for the possibility of mutually supervening but non-identical levels in a system of ontological levels. Suppose that in such a system $\sigma : \Omega \to \Omega'$ and $\sigma' : \Omega' \to \Omega$ are supervenience maps. Then by the definition of a category, the maps are closed under associative composition, so $\sigma \circ \sigma' : \Omega \to \Omega$ is a supervenience map. But because the category is posetal, there can be at most map with domain and codomain Ω, namely the identity 1_{Ω}. Hence $\sigma \circ \sigma' = 1_{\Omega}$. It follows that σ and σ' must be total *and* surjective.[14] Moreover, if σ and σ' are total *and* surjective then $\sigma \circ \sigma' = 1_{\Omega}$. This is as expected: two ontological levels are isomorphic in a system of levels if and only if they supervene on the other as a whole. So if a supervenience map is ever identified as non-surjective (or partial), the levels it relates cannot be isomorphic.

Like with the first extension to partial maps, these insights about systems of ontological levels apply equally to systems of levels of description, although the application to systems of levels of grain requires a bit more work—it also requires adopting the first extension. To do this, one must consider the partition at each level to include one special equivalence class, the class of worlds *not included at that level*. Then, if two levels of grain are related by a supervenience map, that map must exclude the equivalence class of worlds not included at the level of its domain within its domain of definition. (Of course, other equivalence classes could fall outside its domain of definition, too.) Moreover, instead of requiring the domain of the map to be at least as fine-grained as the codomain, one only requires that the domain of definition be at least as fine-grained as the image of the map. Outside the image, the elements of the codomain need bear no relationship of refinement to those of the domain.

---

[13] For a philosophical investigation of the notion of geometric possibility and its role in the relationism-substantivalism debate, see Belot [7].

[14] If σ were not total, then 1_{Ω} could not be; if σ' were not surjective, then 1_{Ω} could not be. The same reasoning applies mutatis mutandis to $\sigma' \circ \sigma = 1_{\Omega'}$. Hence, each of σ and σ' is both total and surjective.

6.4 Levels as Categories

Another way of extending List's framework is to give a richer account of the internal structure of levels of description. In particular, it can be fruitful to view levels of description as categories of structures.[15] As we will see below, this account is more suitable to capture important scientific levels of description and it establishes a close connection between List's work and current developments in philosophy of science.

More precisely, our proposal is to represent a level of description, L, as a pair $\langle \mathbf{L}, \Omega \rangle$ consisting of a description language, $\mathbf{L}$, and a category, Ω, of $\mathbf{L}$-structures.[16] This account differs from List's own account of levels of description in two ways.

1. The objects in Ω are $\mathbf{L}$-structures rather than maximally consistent sets of $\mathbf{L}$-sentences. This is not a big difference. But, among other things, it allows one to apply model-theoretic notions (such as homomorphism, embedding, isomorphism, etc.) without further ado.
2. Ω is a category rather than a bare class. So to specify a level of description, one does not only specify its structures but also the morphisms (admissible transformations) between these structures. Which morphisms to choose is in general not determined by the language $\mathbf{L}$ alone. The choice of morphisms in Ω is linked to an interpretive choice concerning the description language. It reflects which expressions of $\mathbf{L}$ are taken to be meaningful within the level L. The idea is that an $\mathbf{L}$-expression is only meaningful within L if its extension is invariant under the morphisms in Ω.

A major advantage of this account is that it enables us to deal with levels of description on which there is a non-trivial distinction between meaningful and non-meaningful expressions. Levels of description of this kind are prevalent both in the natural and social sciences because scientific languages often contain auxiliary vocabulary or use numerical descriptions that may be transformed according to certain rules without changing in content.

Let us consider an example from social choice theory: descriptions of individual welfare along the lines of Sen [31]. In contrast to Arrow's ordinal framework [3], Sen's framework rests on a numerical description of the welfare of individuals under given alternatives. We reconstruct a level of description of welfare $L_{\mathrm{wf}} = \langle \mathbf{L}_{\mathrm{wf}}, \Omega_{\mathrm{wf}} \rangle$ along these lines to illustrate the proposed account.

The language $\mathbf{L}_{\mathrm{wf}}$ comprises the following descriptive symbols: (1) a sort symbol I for individuals, (2) a sort symbol A for alternatives, and (3) a function symbol W of type $I \times A \rightarrow \mathbb{R}$. If i and a are terms of the sorts I and A, respectively, then $W_i(a)$ is a term of sort $\mathbb{R}$. It stands for the degree of i's welfare under alternative a. Of course, the language also has mathematical auxiliary symbols such as $\mathbb{R}$, $+$, $\cdot$, $<$, etc. Moreover, we assume that the language has the usual logical vocabulary (connectives, quantifiers, and variables) and formation rules for formulas.

---

[15] So on this account, systems of levels of description can be viewed as categories of categories.

[16] We use the term "**L**-structure" in its usual model-theoretic sense. So an **L**-structure is an assignment of extensions to the descriptive symbols of **L**.

An object in Ω_{wf} is an $\mathbf{L}_{\mathrm{wf}}$-structure that consists essentially of a finite set of individuals, a finite set of alternatives and a welfare profile for these individuals and alternatives. To put it more rigorously, the objects in Ω_{wf} are those $\mathbf{L}_{\mathrm{wf}}$-structures, ω, such that (a) ω expands the standard model of the auxiliary mathematical part of the vocabulary (including the field of real numbers), and (b) I^{ω} and A^{ω}—the sets associated to I and A in the structure ω—are both finite.

Which morphisms to include in Ω_{wf} depends on which expressions of $\mathbf{L}_{\mathrm{wf}}$ one considers to be meaningful. List [23] gives a concise overview of various positions, each of which depends on the extent to which interpersonal comparisons of welfare are considered as meaningful and on which level of measurement (ordinal, interval, or ratio) welfare data are taken to reside. For instance, many economists (including Arrow) consider only ordinal *intra*personal comparisons of welfare as meaningful.[17] According to this view, the formula "$W_i(a) < W_i(a)$" is meaningful. In contrast, the formula "$W_i(a) < W_j(a) \wedge i \neq j$" is not meaningful because it amounts to an *inter*personal comparison of welfare. And "$W_i(a) = 2 \cdot W_i(b)$" is also not meaningful because it goes beyond a comparison of order. On the other end of the spectrum, one finds the view that full interpersonal comparisons are meaningful and welfare can be measured on a ratio scale. According to this view, both "$W_i(a) < W_j(a) \wedge i \neq j$" and "$W_i(a) = 2 \cdot W_i(b)$" are meaningful.

Much depends on which descriptions are taken as meaningful. A level of numerical welfare descriptions where only ordinal intrapersonal comparisons count as meaningful boils down to a purely ordinal level of description of welfare along the lines of Arrow [3]. The meaningful numerical statements of the form "$W_i(a) < W_i(b)$" can be reduced to statements about individual preference orderings of the form "individual i prefers alternative a to alternative b" (or, in short: "$a \prec_i b$"). In contrast, a level of numerical welfare descriptions where interpersonal ratio comparisons of welfare count as meaningful is an essentially more fine-grained level of description. Thus, even if levels of description have the same underlying language and structures, they are not necessarily the same. One also needs to specify what counts as meaningful to identify a level.

As alluded to above, different standards of meaningfulness can be captured in terms of choices of morphisms. To discuss such choices in our present example, the notion of a transformation of structures is useful. Given structures $\omega_1, \omega_2 \in \Omega_{\mathrm{wf}}$ with the same set of individuals $\mathrm{Ind} := I^{\omega_1} = I^{\omega_2}$ and the same set of alternatives $\mathrm{Alt} := A^{\omega_1} = A^{\omega_2}$, we call a family $\{f_i\}_{i \in \mathrm{Ind}}$ of functions from $\mathbb{R}$ to $\mathbb{R}$ a *transformation of* ω_1 into ω_2 just in case for all $i \in \mathrm{Ind}$,

$$f_i \circ W_i^{\omega_1} = W_i^{\omega_2}.$$

Now the question of which welfare comparisons one considers to be meaningful boils down to the question of which transformations one considers to be admissible in the

[17]For a discussion of the thesis that interpersonal comparisons of utility are meaningless and the inference that such comparisons are therefore impossible, see List [22].

sense that they do not change any meaningful features of welfare profile structures. The admissible transformation are then taken as morphisms in the category.

To capture the idea that only ordinal intrapersonal comparisons of welfare are meaningful, one takes all positive monotonic transformations as morphisms (cf. Sen [31, Chap. 7*]). More precisely, one takes $\{f_i\}_{i \in \text{Ind}}$ as a morphism from ω_1 to ω_2 just in case it is a transformation of ω_1 into ω_2 such that, for all $i \in \text{Ind}$, f_i is a positive monotonic function. To capture the ideas that full interpersonal and ratio comparisons are meaningful, one takes $\{f_i\}_{i \in \text{Ind}}$ as a morphism from ω_1 to ω_2 just in case it is a transformation of ω_1 into ω_2 such that for some $\alpha > 0$, $f_i(x) = \alpha \cdot x$ for all $x \in \mathbb{R}$ and all $i \in \text{Ind}$. Let $\Omega_{\text{wf}}^{\text{ONC}}$ be the category with the former choice of morphisms and $\Omega_{\text{wf}}^{\text{RFC}}$ the category with the latter.[18] Then, according to our proposal, $\langle \mathbf{L}_{\text{wf}}, \Omega_{\text{wf}}^{\text{ONC}} \rangle$ and $\langle \mathbf{L}_{\text{wf}}, \Omega_{\text{wf}}^{\text{RFC}} \rangle$ are different levels of description. This illustrates how taking morphisms into account allows us to capture differences between levels of description even if they have the same underlying language and structures.

Let us now turn to supervenience in the extended framework. Since levels of description are treated as categories of structures, supervenience relations between them are best viewed as *functors*. Note that any functor maps isomorphic objects to isomorphic objects.[19] This coheres with the idea of global supervenience, that sameness of worlds on the subvening level implies sameness of worlds on the supervening level.

But more importantly, viewing supervenience relations as functors allows us to shed more light on List's requirement of surjectivity. First, it is important to distinguish between surjectivity and essential surjectivity. A functor $F : \Omega \to \widetilde{\Omega}$ is *essentially surjective* if and only if every object $\widetilde{\omega}$ in $\widetilde{\Omega}$ is isomorphic to $F(\omega)$ for some object ω in Ω. From a category-theoretic point of view, essential surjectivity is a fruitful notion. In contrast, surjectivity simpliciter is much too strict. If a level of description is such that every structure has several representationally equivalent variants, there is no good reason to require that a supervenience map from a lower level to this level be surjective. Requiring this would preclude cases of supervenience where a level on which every structure has several representationally equivalent variants supervenes on a level without (an at least equal number of) these variants. A given structure ω at the subvening level would correspond to many representationally equivalent structures at the supervening level. However, a supervenience map can send ω only to one of those. So the others cannot be in the image of the map. Ruling out such cases would be a serious limitation. It would, for instance, make it impossible to capture cases in which non-quantitative descriptions determine corresponding quantitative descriptions up to certain transformations. But many important cases of supervenience between levels of description are of this type, especially those

---

[18]"ONC" stands for "ordinal measurability with no interpersonal comparability" and "RFC" is short for "ratio-scale measurability with full interpersonal comparability." These acronyms are due to List [23].

[19]If F is a functor from the category Ω to the category Ω' and ω_1 is isomorphic to ω_2 in Ω, then $F(\omega_1)$ is isomorphic to $F(\omega_2)$ in Ω'.

given by theorems in measurement theory that exhibit the existence of numerical representations and their uniqueness up to certain transformations [21, 33, 34].

This suggests that surjectivity has to be given up regardless of the arguments given in Sect. 6.3. So even if one would like to adhere to the idea that every world on the supervening level must have a corresponding world on the subvening level, one should better explicate this idea in terms of essential surjectivity rather than surjectivity simpliciter. In view of that, it becomes clear that the arguments in Sect. 6.3 are in fact arguments in favor of dropping even essential surjectivity.

And it makes sense to go even further. There is no principled reason why a system of levels should only incorporate supervenience functors and no other functors between levels. From the perspective of the proposed extension of List's account, it makes sense to include *all functors* between levels of description as arrows in the category of system of levels. Then, rather than excluding and thereby neglecting many functors from the start, one may explore and classify the entire zoo of functors between levels. For example, one may ask which functors should count as reduction functors and which should be seen merely as supervenience functors. Or by dropping uniqueness of arrows between levels, one can investigate cases of multiple reducibility: that is, different ways of reducing a higher level to a lower level (analogous to different ways of reducing arithmetic to set theory).[20]

Here is an example to illustrate how this could look like. Political scientists describe individual preferences on different levels. On one level, one can describe each individual's preference scores for the alternatives in question. On another level, one can describe which alternatives each individual approves of. The structures of the former level are all logically possible preference profile scores for a set of alternatives, and a set of individuals. The structures of the latter level are all possible approval profiles for such sets. Now, one can reduce the level of approval descriptions to the level of preference score descriptions by setting a threshold t such that all alternatives with a preference score above t are taken as approved of. Thus every preference score profile gives rise to an associated approval profile. Relative to this supervenience map, statements of the form "individual i approves of alternative a" reduce to statements of the form "individual i's preference score of alternative a is above t". The crucial point is that there are different choices of thresholds that succeed equally well at reconstructing all approval profiles from preference score profiles. So there are multiple reductions of the level of approval descriptions to the level of preference score descriptions. However, a given approval profile will in general correspond to different preference score profiles under different reductions. So there are substantial differences between such reductions. But it is not clear a priori which reductions are "better" than others and in which respects. In such cases, it

[20]One might seek to model multiple realizability in this way: for example, perhaps one successful reduction translates "pain" by "firing of C-fibres" whilst another translates "pain" by "firing of D-fibres" (where, let us suppose, human brains have C-fibres and Martian brains have D-fibres). However, it is not clear to us what the prospects for this manoeuvre might be; note that neither translation will map the true higher-level claim "both humans and Martians experience pain" to a true lower-level claim.

makes sense to investigate a variety of possible reductions within a system of levels of description.

To sum up, we propose to generalize List's framework in two ways: (1) we construe levels of description as categories rather than bare classes, and supervenience relations as functors; (2) we allow all sorts of functors to be included in a system of levels of description. An advantage of this radical generalization is that it makes the rich toolbox of category-theoretic concepts available for the analysis of levels of description and their relations. For example, the concepts of natural transformation, equivalence, duality, adjunction, and forgetful functor lend themselves well to this endeavour.

This point also plays an essential role in a new strand of research in philosophy of science which uses these category-theoretic concepts to study relations between physical theories (where scientific theories are represented as categories of structures).[21] The proposed generalization of List's framework establishes a close connection to this new work in philosophy of science. Many results from this literature can be viewed as results about how certain types of physical descriptions are related to each other and, thus, how our current system of levels of physical descriptions is structured. But the fruitfulness of the generalized framework is not limited to physics or the natural sciences. As our examples above indicate, the same methods may be applied to levels of description that belong to the social sciences (e.g., economics or political science). We believe that extending and generalizing List's framework makes it better applicable to and, thus, more relevant for philosophy of science in general.

Let us illustrate how the above-mentioned category-theoretic concepts can be used to analyze inter-level relations. As pointed out above, welfare can be described on a quantitative level (in terms of real-valued welfare functions) and on a qualitative level (in terms of mere preference orderings). Let $\langle \mathbf{L}_{\mathrm{ord}}, \Omega_{\mathrm{ord}} \rangle$ be the ordinal level of description. Each object of Ω_{ord} is a structure consisting of a finite set of individuals, a finite set of alternatives, and an assignment of a weak ordering over the alternatives to each individual. To reflect that the identities of alternatives and individuals are in general significant, Ω_{ord} contains only identity morphisms. Suppose the ordinal level $\langle \mathbf{L}_{\mathrm{ord}}, \Omega_{\mathrm{ord}} \rangle$ supervenes on the quantitative level $\langle \mathbf{L}_{\mathrm{wf}}, \Omega_{\mathrm{wf}}^{\mathrm{RFC}} \rangle$ via a supervenience functor F that maps every welfare profile ω in $\Omega_{\mathrm{wf}}^{\mathrm{RFC}}$ to the induced profile of preference orderings ω' in Ω_{ord} and all morphisms between objects ω_1 and ω_2 in $\Omega_{\mathrm{wf}}^{\mathrm{RFC}}$ to the identity morphism on $F(\omega_1) = F(\omega_2)$ in Ω_{ord}. Then there is a sense in which qualitative descriptions neglect or "forget" some information available on the quantitative level. This informal idea is captured formally by the fact that the supervenience functor F is what is called a *forgetful functor* in category theory. In category theory, there is also a way of making precise *what* a functor forgets in terms of its formal properties.[22] In this case, the supervenience functor is essentially sur-

[21] See for example Halvorson and Tsementzis [18] and Hudetz [19].

[22] Baez, Bartels, and Dolan [4] have developed a classification of functors with respect to whether they forget structure, properties, or stuff in terms of whether they are full, faithful, or essentially surjective. For a nice overview and further applications in physics, see Weatherall [36].

jective and also full (i.e., surjective on morphisms). But it is neither faithful (injective on morphisms) nor essentially injective on objects. In technical terms, the functor forgets “stuff”—intuitively speaking: the set of possible degrees of welfare.

But note that this analysis is more subtle than the obvious point that $\langle \mathbf{L}_{\mathrm{wf}}, \Omega_{\mathrm{wf}}^{\mathrm{RFC}} \rangle$ uses numbers while $\langle \mathbf{L}_{\mathrm{ord}}, \Omega_{\mathrm{ord}} \rangle$ does not. That $\langle \mathbf{L}_{\mathrm{wf}}, \Omega_{\mathrm{wf}}^{\mathrm{RFC}} \rangle$ uses numbers is not the point here. The level $\langle \mathbf{L}_{\mathrm{wf}}, \Omega_{\mathrm{wf}}^{\mathrm{ONC}} \rangle$ of numerical welfare descriptions where only ordinal intrapersonal comparisons are taken as meaningful also uses numbers but it should count as equivalent to the purely ordinal level $\langle \mathbf{L}_{\mathrm{ord}}, \Omega_{\mathrm{ord}} \rangle$. And, indeed, their equivalence can be captured in category-theoretic terms using the notion of an equivalence of categories. An *equivalence between categories* C and D is a pair of functors $F : C \rightleftarrows D : G$ that are essentially inverse to each other.[23] That there is an equivalence of categories between $\langle \mathbf{L}_{\mathrm{wf}}, \Omega_{\mathrm{wf}}^{\mathrm{ONC}} \rangle$ and $\langle \mathbf{L}_{\mathrm{ord}}, \Omega_{\mathrm{ord}} \rangle$ can be demonstrated easily by invoking the representation theorem for weak preference orderings. Examples of category-theoretic equivalences abound in the philosophy of physics literature. See, for example, Barrett's work [5] on Hamiltonian and Lagrangian descriptions of classical mechanical systems or Weatherall's work [35, 37] on classical field theories.

Another relation which is naturally captured in category-theoretic terms is that of duality between levels of description. Categories C and D are *dual* to each other if and only if there is an equivalence between C and D^{op}, where D^{op} is the opposite of the category D and it is given by reversing all the morphisms in D. One often find dualities between algebraic and topological/geometric levels of description. For example, Rosenstock, Barrett, and Weatherall [29] show that the usual manifold formulation of general relativity is dual to the algebraic formulation in terms of Einstein algebras in the sense that these theories have dual categories of structures.

Another important category-theoretic concept is that of adjunction. Although this notion is hard to grasp on a pre-theoretic level, it may still be fruitful for analyzing relations between levels of description. In our context, adjunction can be roughly understood as a special type of supervenience where the supervenience functor is accompanied by a second functor in the other direction: its adjoint. As Feintzeig [15] has shown, the concept of adjunction captures the relationship between the level of infinite (limiting) quantum systems and the level of finite quantum systems in quantum statistical mechanics. The basic idea is that the properties of infinite quantum systems are determined in a specific way by the properties of their finite subsystems. This illustrates that the generalized framework allows one to transfer technical concepts to new domains of application where they can be used to capture relations between levels of description for which we would otherwise lack appropriate concepts.

[23] This means that their composition FG is naturally isomorphic to the identity functor on C and GF is naturally isomorphic to the identity functor on D.

6.5 Conclusions and Prospects

In Sect. 6.1, we reviewed List's levels formalism [24], focusing on his account of systems of ontological levels, in particular systems of levels of grain and systems of descriptive levels. One of the results of this review was to qualify his claim that any system of ontological levels can be mapped functorially into some system of levels of grain: this is true only for systems of ontological levels that are downward-directed, i.e., for which each pair of levels has a common subvenient level that supervenes on all such common levels. In Sect. 6.2, we examined the relation between supervenience and reduction in this formalism. We showed that while in general supervenience does not imply reduction, it does so imply it when the levels related by the supervenience map are compatible and jointly characterizable. Compatibility requires, roughly speaking, that levels of description related by a supervenience map agree with each other as far as shared vocabulary is concerned. Joint characterizability relative to a supervenience map is a strong condition. It holds when the union of two levels of description relative to a supervenience map admits of a description itself. But in many cases of supervenience between scientific levels of description, this can be expected. So it is quite plausible that in many cases of interest, supervenience and reduction of levels go hand in hand.

After this analysis, in Sect. 6.3 we proposed two extensions of systems of ontological levels by weakening List's characterization of their supervenience maps as surjective functions. First, we proposed considering merely partial (instead of total) functions, motivated by the idea that not every lower-level world ought to give rise to some higher-level world. The totality of the supervenience maps could be recovered by introducing null worlds at each level, but in many cases such worlds would have to be interpreted with caution, perhaps as just mathematical conveniences. Second, we proposed dropping the requirement of surjectivity. While this requirement is plausible for ontological naturalists, we do not think that it should be encoded into the definition of a system of levels, which, as a formal tool, ought to be propounded as neutrally as is feasible regarding substantive philosophical positions. Moreover, it also fits better with philosophy of science applications, as discussed in Sect. 6.4, in which it is more fruitful to consider a level not as a bare class (of possible worlds), but as a category itself whose objects are structures of some language. This allows one to distinguish levels of description even when they have the same language and the same structures. We illustrated this with an example of two social choice theories whose models were identical but whose isomorphism classes were distinct. Moreover, since one may have isomorphic but distinct models in a level, one might only require that the image of lower-level models under a supervenience map intersect with each isomorphism class of models of the higher level (i.e., the level that is the codomain of the map). These investigations also follow up on a comment made briefly in Sect. 6.1, that List's framework can be well-characterized without the use of category theory, which just provides a compact description and interpretive gloss for it. By contrast, taking levels as categories themselves demands a more robust use of categorial ideas that could also prove to be more fruitful.

It also suggests accordingly several directions for future research, of which we mention two. First, more examples should be formalized within the generalized framework to test its fruitfulness, flexibility, and power. Obvious candidates include levels described by:

- physical theories, such as thermodynamics and statistical mechanics [14], and quantum and classical mechanics [8];
- biological theories, such as classical and molecular genetics, levels of selection, and developmental biology and genetics [10]; and
- theories in political science (and other social sciences) that describe individuals, on the one hand, and groups, on the other [25].

Second, as mentioned in Sect. 6.4, one can generalize the framework of posetal categories to include many types of functorial relationships between levels besides supervenience. List already mentions reduction as one sort that can be defined for systems of levels of description, but the sort of reduction he has in mind (as alluded in Sect. 6.2) follows one of the oldest versions of the Nagelian model for reduction [27], understood as deductibility allowing for definitional extension and bridge laws induced through the supervenience maps. Already in 1967 Schaffner [30] (foreshadowed by Nagel [28] himself) suggested that reduction needs to accommodate the way different theories (here describing different levels) are related by approximations. But supervenience maps seem entirely inapt to capture these. Perhaps there is some more expansive functorial relationship that can capture these notions of reduction that seem more central to science, such as those suggested by recent work on topological (and topologically inspired) structures on models of theories [16].

Acknowledgements ND is primarily responsible for Sects. 6.1 and 6.2. SCF is primarily responsible for Sects. 6.3 and 6.5 and for general editing, and secondarily responsible for Sect. 6.1. LH is primarily responsible for Sect. 6.4 and for Proposition 1, and secondarily responsible for Sect. 6.3 and general editing. All authors thank Katie Robertson for many insightful conversations leading to the genesis of this essay, Tomasz Brengos and Christian List for encouraging comments on a previous version, and the audience and organizers of the workshop "New Perspectives on Inter-Theory Reduction" in Salzburg in November, 2017. SCF acknowledges partial support through a Marie Curie International Incoming Fellowship (PIIF-GA-2013-628533).

References

1. J. Adámek, H. Herrlich, G.E. Strecker, Abstract and concrete categories: the joy of cats, Online edn. (2004), http://katmat.math.uni-bremen.de/acc
2. H. Andréka, J.X. Madarász, I. Németi, Defining new universes in many-sorted logic (2008), https://old.renyi.hu/pub/algebraic-logic/kurzus10/amn-defi.pdf. Unpublished manuscript
3. K. Arrow, *Social Choice and Individual Values* (Wiley, New York, 1951)
4. J. Baez, T. Bartel, J. Dolan, Property, structure, and stuff (2004), http://math.ucr.edu/home/baez/qg-spring2004/discussion.html
5. T.W. Barrett, Equivalent and inequivalent formulations of classical mechanics. Br. J. Philos. Sci. **forthcoming** (2018). https://doi.org/10.1093/bjps/axy017
6. T.W. Barrett, H. Halvorson, Morita equivalence. Rev. Symb. Log. **9**(3), 556–582 (2016)

7. G. Belot, *Geometric Possibility* (Oxford University Press, Oxford, 2011)
8. A. Bokulich, *Reexamining the Quantum-Classical Relation* (Cambridge University Press, Cambridge, 2008)
9. F. Borceux, *Handbook of Categorical Algebra, Vol. 1: Basic Category Theory* (Cambridge University Press, Cambridge, 1994)
10. I. Brigandt, A. Love, Reductionism in biology, in *The Stanford Encyclopedia of Philosophy*, Spring 2017 edn., ed. by E.N. Zalta. Metaphysics Research Lab, Stanford University (2017)
11. T. Button, S. Walsh, Structure and categoricity: determinacy of reference and truth value in the philosophy of mathematics. Philos. Math. **24**(3), 283–307 (2016)
12. J.R.B. Cocketta, S. Lackb, Restriction categories I: categories of partial maps. Theor. Comput. Sci. **270**(1–2), 223–259 (2002)
13. N. Dewar, Supervenience, reduction, and translation. Philos. Sci. **forthcoming** (2018). http://philsci-archive.pitt.edu/15304/
14. G. Emch, C. Liu, *The Logic of Thermostatistical Physics* (Springer, Berlin, 2002)
15. B.H. Feintzeig, Deduction and definability in infinite statistical systems. Synthese **196**(5), 1831–1861 (2019)
16. S.C. Fletcher, Similarity structure on scientific theories, in *Topological Philosophy*, ed. by B. Skowron (de Gruyter, Berlin, forthcoming)
17. H. Halvorson, *The Logic in Philosophy of Science*. (Cambridge University Press, Cambridge, 2019)
18. H. Halvorson, D. Tsementzis, Categories of scientific theories, in *Categories for the Working Philosopher*, ed. by E. Landry (Oxford University Press, Oxford, 2017), pp. 402–429
19. L. Hudetz, Definable categorical equivalence. Philos. Sci. **86**(1), 47–75 (2019)
20. J. Kim, The layered model: metaphysical considerations. Philos. Explor. **5**(1), 2–20 (2002)
21. D.H. Krantz, R.D. Luce, P. Suppes, A. Tversky, *Foundations of Measurement, Vol. II: Additive and Polynomial Representation* (Academic, San Diego, 1989)
22. C. List, Are interpersonal comparisons of utility indeterminate? Erkenntnis **58**(2), 229–260 (2003)
23. C. List, Social choice theory, in *The Stanford Encyclopedia of Philosophy*, Winter 2013 edn., ed. by E.N. Zalta. Metaphysics Research Lab, Stanford University (2013)
24. C. List, Levels: descriptive, explanatory, and ontological. Noûs **forthcoming** (2018). https://doi.org/10.1111/nous.12241
25. C. List, K. Spiekermann, Methodological individualism and holism in political science: a reconciliation. Am. Polit. Sci. Rev. **107**(4), 629–643 (2013)
26. B. McLaughlin, K. Bennett, Supervenience, in *The Stanford Encyclopedia of Philosophy*, Spring 2018 edn., ed. by E.N. Zalta. Metaphysics Research Lab, Stanford University (2018)
27. E. Nagel, The meaning of reduction in the natural sciences, in *Science and Civilization*, ed. by R.C. Stouffer (University of Wisconsin Press, Madison, 1949), pp. 99–135
28. E. Nagel, *The Structure of Science: Problems in the Logic of Scientific Explanation* (Harcourt, Brace & World, New York, 1961)
29. S. Rosenstock, T.W. Barrett, J.O. Weatherall, On Einstein algebras and relativistic spacetimes. Stud. Hist. Philos. Mod. Phys. **52**, 309–316 (2015)
30. K. Schaffner, Approaches to reduction. Philos. Sci. **34**, 137–147 (1967)
31. A.K. Sen, *Collective Choice and Social Welfare* (Holden-Day, San Francisco, 1970)
32. O. Shagrir, Concepts of supervenience revisited. Erkenntnis **78**(2), 469–485 (2013)
33. P. Suppes, *Representation and Invariance of Scientific Structures* (CSLI Publications, Stanford, 2002)
34. P. Suppes, D.H. Krantz, R.D. Luce, A. Tversky, *Foundations of Measurement, Vol. III: Representation, Axiomatization and Invariance* (Academic, San Diego, 1990)
35. J.O. Weatherall, Are Newtonian gravitation and geometrized Newtonian gravitation theoretically equivalent? Erkenntnis **81**(5), 1073–1091 (2016)
36. J.O. Weatherall, Understanding gauge. Philos. Sci. **83**(5), 1039–1049 (2016)
37. J.O. Weatherall, Category theory and the foundations of classical space-time theories, in *Categories for the Working Philosopher*, ed. by E. Landry (Oxford University Press, Oxford, 2017), pp. 329–348

Chapter 7
From Quantum-Mechanical Lattice of Projections to Smooth Structure of $\mathbb{R}^4$

Krzysztof Bielas and Jerzy Król

Abstract Mathematical formalism of quantum mechanics provides an interesting way of thinking not only about real numbers in general, but also about the phenomenon of exotic smoothness. The main point of departure is a set of Boolean algebras contained in the quantum-mechanical lattice of projections, used to build various Boolean-valued models of ZFC, thus "universes for mathematics" in the usual sense. As each of them provides its own notion of real numbers, it gives a rather unique opportunity for considering exotic smoothness questions. It is not a surprise that an important step in the construction is a mapping between Boolean algebras and structures built upon reals, such as manifold covers. The present work aims at a categorical perspective on this subject. In particular, it points at the colimit object in various categories as a principle of combining subobjects into a single structure.

7.1 Introduction

The relation between quantum and classical theories has been undoubtedly a persistent and demanding problem in theoretical physics for decades. Among various approaches that have been made through years, the perspective of quantum reality perceived through "classical glasses", or classical reference frames (as famously adviced by Niels Bohr), seemed to be well-suited to find its rigorous mathematical formulation (cf. [22]). This hope has been driven e.g. by a long-standing research on the interplay between logical structures of classical and quantum theories. While the

K. Bielas (✉) · J. Król
University of Information Technology and Management, ul. Sucharskiego 2, 35-225 Rzeszow, Poland
e-mail: krzysiek.bielas@gmail.com

J. Król
e-mail: iriking@wp.pl

K. Bielas
Institute of Physics, University of Silesia, ul. 75 Pułku Piechoty 1, 41-500 Chorzow, Poland

M. Kuś and B. Skowron (eds.), *Category Theory in Physics, Mathematics, and Philosophy*, Springer Proceedings in Physics 235,
https://doi.org/10.1007/978-3-030-30896-4_7

former comprise mainly Boolean algebras, the latter deal with various nondistributive generalizations thereof, such as orthomodular lattices, effect algebras, orthoalgebras etc. (for an exhaustive literature concerning quantum logics see [31]). Importantly, these can be often "coordinatized" by their classical counterparts, and it is the part of a broader, ongoing subject of determining general algebraic objects, such as C^*-algebras, by their respective commutative substructures [16, 38].

Meanwhile, it has become a widely accepted cosmological scenario that our spacetime started from some initial quantum system S and evolved to a large-scale smooth 4-manifold M. While the choice $M \simeq_{\text{hom}} \mathbb{R}^4$ seems to be the simplest, it also turns out to be the most complicated one from the point of view of smooth structures. In fact, the variety of smooth structures that can be put on $\mathbb{R}^4$ diverges drastically, and one finds uncountably many nondiffeomorphic (exotic) smooth manifolds, each homeomorphic to $\mathbb{R}^4$ [15, 36]. Their physical (including cosmological) significance has been a matter of vast research [1–7, 12, 17, 34]. In particular, let us point to the recent work [25] that merges seemingly distinct pictures of classical and quantum physics as follows.

Let $\mathscr{H}$ be a complex, separable Hilbert space associated to the initial quantum system S. As such, $\mathscr{H}$ carries an orthomodular lattice (OML) $\mathbb{L}$ of projections on closed linear subspaces of $\mathscr{H}$ with a family of its Boolean subalgebras $\{B_i | B_i \subseteq \mathbb{L}\}$. The latter give rise to a family of Boolean-valued models of ZFC, denoted V^B, together with distinct objects of real numbers $\mathbb{R}_B$. Then, assuming the smoothness of spacetime M to originate from S, one gains an opportunity to follow Bohr's philosophy and parametrize spacetime by charts associated to local, Boolean reference frames. An immediate result is that one cannot cover whole manifold with a one-element smooth atlas, just as one cannot cover an OML $\mathbb{L}$ with a single Boolean algebra, unless $\mathbb{L}$ is distributive. Since $M \simeq_{\text{hom}} \mathbb{R}^4$ and standard smooth $\mathbb{R}^4$ is the only one that admits a single-chart atlas, the conclusion is that M has to be exotic smooth (for more detailed, model-theoretical perspective on this construction, cf. [25]).

Presently, we elaborate on the intermediate part of above parametrization, that was not shown explicitly in [25]. In more detail, we propose the category theory to be a guiding principle in recognizing suitable structures on the "quantum" side (an OML $\mathbb{L}$ and its Boolean subalgebras) and corresponding ones on the "classical" side (an atlas of a spacetime manifold M). Also, an appropriate map between said domains seems to be best identified in the categorical setup. The present work may be considered as a point of departure for further analysis; more thorough study will be given in separate work [9].

7.2 Quantum Mechanics—Preliminaries

To begin with, let S be a quantum-mechanical system and $\mathscr{H}$ its associated Hilbert space of states. Then, the observables are represented by self-adjoint elements of a C^*-algebra $\mathscr{B}(\mathscr{H})$ of bounded operators on $\mathscr{H}$. One could also start with an abstract C^*-algebra A and then define states of S as positive, normed linear functionals on

A. The equivalence of the approaches is provided through the celebrated Gleasons's theorem which sets the one-to-one correspondence between states $\omega : A \to \mathbb{C}$ and density operators (i.e. positive trace class operators of trace one) $\rho_\omega \in A$ by

$$\omega(a) = \mathrm{tr}(\rho_\omega a) \text{ for any } a \in A, \tag{7.1}$$

as long as $\dim(\mathscr{H}) > 2$ [11]). Further standard axiomatics for empirical content of quantum mechanics such as Born rule etc. will not bother us here (but see e.g. [11] for a proper background).

It is well known that one of the main features of quantum mechanics is that the observables do not commute in general, in contrast with the content of classical theories. This is obviously reflected in the type of algebraic structures at hand in either case. Accordingly, a great deal of research has been devoted to the question of to what extent quantum entities, such as C^*-algebras, von Neumann algebras, or OML's, are determined by their respective classical (i.e. commutative) subobjects, and what are their interrelations. This subject has been attractive from the point of view of both pure mathematics and physics. For example, the Gelfand duality states that there is a duality between the categories of commutative C^*-algebras with *-morphisms and compact Hausdorff spaces with continuous maps. Then, the Gelfand spectrum $\Sigma(A)$ for a general (possibly noncommutative) C^*-algebra A is to be considered as a prototype of a quantum generalized space [21].

In physics, the case of OML's led to the discussion on how this situation resembles local reference frames in special relativity [8, 14] or its possible role in quantization of gravity [14]. In the following we motivate the choice of OML's, rather than general operator algebras, to be appropriate representatives of quantum systems.

Among the observables there is a class of operators of special importance: these are the projection operators (projections) (recall that $p : \mathscr{H} \to \mathscr{H}$ is a projection whenever $p^2 = p = p^\dagger$). The significance of projections in the formulation of a quantum theory lies in the fact that every observable $a : \mathscr{H} \to \mathscr{H}$ can be recovered from them by the so-called spectral theorem [37]

Theorem 1 *For every family $\{a_i\}_{i\in I}$ of self-adjoint pairwise commuting operators, there exists a complete Boolean algebra of projections B such that given the spectral decompositions of each a_i*

$$a_i = \int \lambda \mathrm{d}e^i_\lambda,$$

it holds that $\forall i \in I(\mathrm{d}e^i_\lambda \in B)$

(more on the Boolean algebras of projections one finds in the following).

At the same time, projections can be considered as yes-no propositions about the system with an associated $\mathscr{H}$ and, being in one-to-one correspondence with closed subspaces of $\mathscr{H}$, they form a complete orthomodular lattice

$$\mathbb{L} \equiv (\mathbb{L}, \vee, \wedge, \neg, 0, 1) \tag{7.2}$$

with join, meet and complement defined by

$$\bigvee p_i = p_M \text{ where } M \text{ is the closed linear span of } \bigcup \text{ran}(p_i),$$
$$\bigwedge p_i = p_M \text{ where } M = \bigcap \text{ran}(p_i),$$
$$\neg p = 1 - p,$$

and 0, 1 elements as zero and identity operators on $\mathscr{H}$, respectively [8]. Unsurprisingly, the above operations are to be understood as logical connectives "and", "or" and "not" together with values 0—false and 1—true. The structure of $\mathbb{L}$ appeared for the first time in the context of quantum mechanics thanks to the seminal work of Birkhoff and von Neumann [10]. This way, it gave rise to the first incarnation of the so-called quantum logic, now considered to be a standard choice among other logics for quantum theory that have been proposed since then, cf. [31].

For the sake of clarity, let us mark the main differences that the logic of a classical theory exhibits with the above. It is perfectly valid and general to think of any classical system with configuration space X in terms of its phase space (more precisely, a cotangent bundle T^*X) together with yes-no proposition represented by characteristic functions $\{\chi_U\}$ (projections) of measurable sets $U \subseteq T^*X$. The one-to-one correspondence between $\{\chi_U\}$ and $U \subseteq T^*X$ provides that everything comes under the rules of classical logic, i.e. that of a Boolean algebras. On the other hand, it is well known that the quantum lattice $\mathbb{L}$ is not Boolean whenever $\dim(\mathscr{H}) > 2$ (for a simple illustration of non-distributivity, consider $\dim(\mathscr{H}) = 3$ and three mutually orthogonal projections).

7.3 Boolean Subalgebras of $\mathbb{L}$

In all its quantumness, $\mathbb{L}$ contains plethora of Boolean algebras, although none of them covers whole $\mathbb{L}$ at once (recall that $B \subseteq \mathbb{L}$ is a Boolean subalgebra of $\mathbb{L}$ whenever $B \equiv (B, 0, 1, \vee|_B, \wedge|_B, \neg|_B)$ is a Boolean algebra). Let BSub($\mathbb{L}$) denote the set of all Boolean subalgebras of $\mathbb{L}$, partially ordered by inclusion. A trivial example of an element of BSub($\mathbb{L}$) is $B_2 = \{0, 1\}$. Another class of examples would be the family of $B_p = \{0, p, \neg p, 1\}$, $p \in \mathbb{L}$ which shows that every $p \in \mathbb{L}$ is contained in at least one element of BSub($\mathbb{L}$). A somewhat more involved example would be an atomless Boolean algebra B_Q containing the spectral family of a position operator Q. By Zorn's lemma it follows that every $B \in$ BSub($\mathbb{L}$) is contained in some (not necessarily unique) maximal $B' \in$ BSub($\mathbb{L}$); such maximal elements of BSub($\mathbb{L}$) are called blocks and $\mathbb{L}$ can be decomposed into them in an obvious way [35].

Let us now formulate explicitly what we mean by "Boolean reference frames" throughout the paper; the language of homomorphisms between appropriate logical structures seem to be the most instructive here. Recall that one of the versions of the Kochen–Specker theorem claims that whenever $\dim(\mathscr{H}) > 2$, then $\mathbb{L}$ admits no global truth-valuation, i.e. there is no homomorphism $\mathbb{L} \to B_2$ [32]. This is to be understood as the impossibility to give a truth-valuation to all propositions that can be made about a general quantum system [24]; in fact, the existence of such a global valuation $\omega : B \to B_2$ is equivalent to ω being a dispersion-free on B (i.e. $\forall p \in B(\omega(p^2) = \omega(p)^2)$ [13]). On the other hand, dispersion-free states on B are the ones that posess determined value on any observable with a spectral decomposition in B (see Theorem 1). This way, each Boolean subalgebra $B \in \mathrm{BSub}(\mathbb{L})$ is to be considered as a local, classical frame of reference for a quantum system with the associated lattice $\mathbb{L}$.

The inverse problem of a *block decomposition* answers the question, whether given a family $\mathscr{B}$ of Boolean algebras, they give rise to an OML. The usual way to combine the elements of $\mathscr{B}$ is called the *pasting* of $\mathscr{B}$ [23, 35]; this can be done in more than one way, the most primitive one being the $\{0, 1\}$-pasting, i.e. gluing the elements of $\mathscr{B}$ by identifying their least and greatest elements.

Now, in the category-theoretic context we would like to identify appropriate category to speak about any OML and its Boolean subalgebras. Let **OML** be a category of orthomodular lattices as objects and lattice homomorphisms as arrows and let $\mathrm{BSub}(\mathbb{L})$ be defined as above; trivially $\mathrm{BSub}(\mathbb{L})$ turns into poset subcategory (every Boolean algebra is a distributive orthomodular lattice at first). The first goal would be to combine the elements of $\mathrm{BSub}(\mathbb{L})$ into a larger entity, presumably the $\mathbb{L}$ itself. In category theory, this happens often by means of a (co)product, or more generally by a (co)limit. In the context of **OML** it proceeds as follows. Let

$$\mathbf{D} = \{B \hookrightarrow \mathbb{L} | B \in \mathrm{BSub}(\mathbb{L})\} \tag{7.3}$$

be a diagram in **OML** that consists of inclusion arrows into $\mathbb{L}$. This makes **D** a cocone in **OML**; a limit for **D** can be defined to be a terminal object in the category of cocones over **D**.

Here it is perhaps worth mentioning that one of the weak points of **OML** indicated frequently [33] is the lack of tensor products, namely given $L_1, L_2 \in Ob(\mathbf{OML})$ it does not necessarily hold that $L_1 \otimes L_2 \in Ob(\mathbf{OML})$. This is the consequence of the fact that the category **OML** is not (co)complete.

Fortunately, following [38] it is possible to extend **OML** to a category **PBool** of *partial* Boolean algebras as objects and appropriate homomorphisms as arrows. It is not necessary to include the exact definition for **PBool** for the purpose of the present work. Let us summarize this by two statements: first, objects in **PBool** are "partial algebras" B with a Boolean algebra structure defined only "locally", i.e. on proper subsets of B (these are then called the total subalgebras). Secondly, **PBool** is a slight generalization of **OML** (in particular, every object of **OML** is an object of **PBool** and **OML** is a subcategory of **PBool**). Moreover, as **PBool** is (co)complete, one goes to all (co)limits freely (particularly in the case of (7.3)) and obtains the following [38].

Theorem 2 *Every partial Boolean algebra is a colimit of its (finitely generated) total subalgebras.*

Applying this to our specific case, we state the following

Corollary 1 *Every orthomodular lattice is a colimit of its (finitely generated) Boolean subalgebras (in the category* **PBool***).*

Let us remark that this is a nice categorical generalization of the result of Harding [20] stating that every orthomodular lattice is determined up to isomorphisms by its semilattice of Boolean subalgebras BSub($\mathbb{L}$). Since (co)limits are always unique up to an isomorphism in any category by definition, the above result follows easily from Corollary 1.

Once again, let us remind that this class of results singles out the algebras for logic of a quantum theory (i.e. those with logical operations defined upon, e.g. orthomodular lattices, partial Boolean algebras etc.) . On the opposite, general C^*-algebras or von Neumann algebras cannot be determined by their commutative subalgebras; this can be done e.g. only up to a commutator (or the so-called Jordan structure), cf. [16, 29].

Let us now turn to the case of classical physics, namely the smooth spacetime structure, to be formulated in the language of differential geometry.

7.4 The Smooth Structure of a Spacetime

We have seen already that an orthomodular lattice of projections $\mathbb{L}$ arises in the context of quantum mechanics as a colimit of its Boolean subalgebras in the category **PBool**. In the following we will try to transfer the structure into the description of spacetime as a differentiable manifold. Presumably, having the category **PBool** on the quantum side, one could proceed "classically" in the following way. Consider some broad category such as the category n-**Mfd** of smooth n-manifolds and smooth maps between them. Although these lack many categorical constructs, including general colimits, we will see that every object in n-**Mfd** can be decomposed analogically to the quantum case of an OML $\mathbb{L}$.

Let us start with a brief reminder of basic definitions [30], in order to grasp its categorical reformulation better. Therefore, we say $(X, \{U_i, \phi_i\})$ is an n-dimensional differentiable (smooth) manifold if $(X, \{U_i, \phi_i\})$ is an n-dimensional topological manifold with an atlas $\{U_i\}$ (i.e. $\{U_i\}$ cover X and each $\phi : U_i \to \mathbb{R}^n$ is a homeomorphism) and each $\phi_i \circ \phi_j^{-1}$, whenever defined, is smooth. Equivalently, one can think of a smooth manifold M as follows: given an atlas $\{f_i : M \supseteq U_i \to V_i \subseteq \mathbb{R}^n\}$ it holds that

$$M = \bigsqcup V_i / \sim \tag{7.4}$$

where $f_i(p) \sim f_j(p)$ whenever $p \in U_i \cap U_j$. This is especially well-suited form for further investigation.

Recall that in the previous section it was described how to construct an orthomodular lattice from its Boolean subalgebras, and the procedure was identified through a colimit object in **PBool**. We have already discussed above that, from the physics viewpoint, Boolean subalgebras provide the system of classical reference frames for a given quantum system, and this can be argued on various levels. Crucially, if the quantum system is to be understood as the initial singularity S, a first guess would be that formally S may give rise to a differential structure of spacetime M. Then a rather natural question is whether it is possible to transfer the colimit in **PBool** to appropriate colimit by means of M.

To identify category-theoretic construction that lurks behind a smooth manifold, consider the following illustrative example [28], to be generalized afterwards. Let $M = S^2$ and observe that M may be viewed in a two-fold way: the first would be through an equalizer

$$S^2 \hookrightarrow \mathbb{R}^3 \underset{t}{\overset{s}{\rightrightarrows}} \mathbb{R}$$

where

$$s(x, y, z) = x^2 + y^2 + z^2 \quad \text{and} \quad t(x, y, z) = 1 \tag{7.5}$$

(the sphere equation $x^2 + y^2 + z^2 = 1$ is captured by the commutativity of the above diagram). The second approach to parametrize S^2 would be through a coequalizer

$$S^1 \times (0, 1) \rightrightarrows D \sqcup D \longrightarrow S^2$$

where D is a two-dimensional disk (the quotient by the glued part of $D \cup D$ is provided again by the commutativity of the above diagram).

To stay independent of any coordinate system, while keeping in mind that we would like to parametrize the general manifold by respective subobjects, we will stick to the second picture, and ask if there is an appropriate construction for any atlas in general. The answer is not only positive; it agrees with the object met earlier in the context of an OML, i.e. that of a pasting [27]:

Theorem 3 *Let* $\mathbf{U_n}$ *be a subcategory of* n-**Mfd** *consisting of all open subsets of* $\mathbb{R}^n$ *and smooth maps between them. Then every object in* n-**Mfd** *is a pasting of objects from* $\mathbf{U_n}$.

Again, transferring the notion of a pasting into category theory and generalizing the coequalizer diagram of S^2, we obtain the following [26].

Corollary 2 *Any object* M *in* n-**Mfd** *is a coequalizer*

$$\bigsqcup W_{ij} \rightrightarrows \bigsqcup W_i \longrightarrow M$$

where $W_i = f_i(V_i)$ *and* $W_{ij} = f_i(V_i \cap V_j)$.

Therefore, we arrived at the following conclusion: given a smooth manifold, it is always the colimit of its atlas.

7.5 Correspondence: From $\mathbb{L}$ to Smooth Structures

Finally, an interesting question is how to relate above categorical structures on both quantum and classical spacetime side. Firstly, given a quantum-mechanical system S with an associated Hilbert space $\mathscr{H}$ and its OML $\mathbb{L}$, suppose we demand for each $B \in \mathrm{BSub}(\mathbb{L})$ a corresponding atlas element U_B of a spacetime M that originated from S. As one cannot cover an OML $\mathbb{L}$ with a single Boolean algebra structure in the general case, one also will not cover whole M with a single chart, and we obtain the following [25].

Corollary 3 *With the above correspondence* $\mathrm{BSub}(\mathbb{L}) \ni B \mapsto U_B \subseteq M$ *and assuming* $M \simeq_{\mathrm{hom}} \mathbb{R}^n$, *we conclude that* M *has to be exotic smooth and* $n = 4$.

Now, it is interesting whether the colimit correspondence is of functorial character, i.e. to what extent one can assign to Boolean frames of reference (i.e. objects in $\mathrm{BSub}(\mathbb{L})$ and trivially in **PBool**) open subsets of $\mathbb{R}^n$ (i.e. objects in $\mathbf{U_n}$ and trivially in n-**Mfd**), such that the assignment obeys the rules of a (covariant) functor $F :$ **PBool** $\to n$-**Mfd**:

$$F(B) = U_B \in \mathrm{Ob}(\mathbf{U_n}) \subseteq \mathrm{Ob}(n\text{-}\mathbf{Mfd}) \quad \text{and} \quad F(f) \in \mathrm{Arr}(n\text{-}\mathbf{Mfd}) \tag{7.6}$$

for any $B \in \mathrm{Ob}(\mathbf{PBool})$ and $f \in \mathrm{Arr}(\mathbf{PBool})$, together with $F(f \circ g) = F(f) \circ F(g)$. (Note that in the extension of this setting, global automorphisms of $\mathbb{L}$ should correspond to diffeomorphisms of the whole M.) We point out that the existence of such a functor F is an open issue left for future research [9]; nevertheless we indicate two points that may be important in the construction thereof.

First, let us recall that $\mathrm{BSub}(\mathbb{L})$ is not only a mere posetal category; it constitutes also a meet-semilattice. In other words, for any $B_1, B_2 \in \mathrm{BSub}(\mathbb{L})$ we have $B_1 \wedge B_2 := B_1 \cap B_2 \in \mathrm{BSub}(\mathbb{L})$. In category-theoretic terms, that is to say that the pullback

$$\begin{array}{ccc} B_1 \cap B_2 & \hookrightarrow & B_2 \\ \downarrow & & \downarrow \\ B_1 & \hookrightarrow & \mathbb{L} \end{array}$$

always exists. For an analogous result in the context of smooth manifolds, we have to go to a subcategory **para**-n-**Mfd** of *paracompact* manifolds (in fact, this is the usual category to talk about spacetime in the standard approach). Then, one can show that each object in **para**-n-**Mfd** admits a good open cover, i.e. given any $X \in \mathrm{Ob}(\mathbf{para}\text{-}n\text{-}\mathbf{Mfd})$, all its chart intersections are diffeomorphic to $\mathbb{R}^n$. Therefore, under the functor $F|_{\mathbf{para}\text{-}n\text{-}\mathbf{Mfd}}$ an intersection of elements of $\mathrm{BSub}(\mathbb{L})$ should correspond again to an object in $\mathbf{U_n}$.

Second, let us consider [39] with the construction of adjoint functors

$$\mathbf{L} : \mathbf{Set}^{[\mathrm{Bor}(\mathbb{R})/B]^{op}} \underset{\longrightarrow}{\longleftarrow} [\mathrm{Bor}(\mathbb{R})/L] : \mathbf{R}$$

This gives an opportunity of studying a "quantum" comma category $[\mathrm{Bor}(\mathbb{R})/L]$ in terms of presheaves on a "classical" $[\mathrm{Bor}(\mathbb{R})/B]$ and vice versa. As the objects of $[\mathrm{Bor}(\mathbb{R})/L]$ are obviously the arrows of **PBool**, it seems reasonable to look for analogous adjoint in the category n-**Mfd** [9].

7.6 Discussion

This work puts some initial remarks on the possibility that the smooth structure of a manifold X can be studied through a category-theoretic correspondence between X, together with its atlas of charts, and some quantum (i.e. noncommutative) object parametrized by its classical (i.e. commutative) subobjects. The first and immediate conclusion is that, when applied to the physical choice $X \simeq_{\mathrm{hom}} \mathbb{R}^n$, it results in exotic smooth structure of X, and therefore its four-dimensionality. Thus, the procedure accounts for the fact that globally we parametrize our spacetime with four variables: one time t and three space directions x, y, z. Moreover, it should be emphasized that this happens purely on the grounds of quantum-mechanical view of the initial state of the Universe.

One should make a distinction about the cardinality of the smooth atlas of exotic R^4. In the case of small R^4 (i.e. embeddable in S^4) there always appear Casson handles—infinite geometric constructions with the topological interior homeomorphic to $\mathbb{R}^4$. Any stage $n \in \mathbb{N}$, representing the level of Casson handle, corresponds to a 3-submanifold Y_n of R^4 defining the sequence $\{Y_n\}_{n\in\mathbb{N}}$ of 3-submanifolds. This sequence defines a highly nontrivial foliation of the entire R^4 (see e.g. [1]). In particular, the C^*-algebra of tensor densities of the foliation contains the von Neuman factor III_1. On the other side, the foliation and the sequence $\{Y_n\}_{n\in\mathbb{N}}$ are responsible for the infinite atlases of small exotic R^4's. Namely, there does not exist any time coordinate in any Y_n that could be properly extended over Y_{n+1}, hence R^4 needs at least countably infinitely many charts in the atlas. As Freedman and Taylor showed [19] there exists the universal exotic R^4 into which any other exotic R^4 embeds. This would indicate the necessity of considering also uncountable atlasses. Thus covering the universal R^4 by uncountable atlas generated from $\mathbb{L}$ defines, in principle, covers of other exotic R^4's as submanifolds. Accordingly, it agrees with the general remark on the cardinality of $\mathrm{BSub}(\mathbb{L})$: whenever $\dim(\mathscr{H}) = \infty$, there is uncountably many noncommuting projections on $\mathscr{H}$, hence also uncountably many blocks in $\mathbb{L}$. However, there remains the case of an exotic R^4 with a finite atlas, which is not covered by this approach. In fact, we do not know whether such R^4's exist at all. If positive, there would exist an exotic 4-disk and exotic S^4 as well. At the same time, it is equivalent to the negation of the (unresolved so far) 4-dimensional smooth Poincaré conjecture. Thus the case of finite atlases of exotic R^4's needs some other tools and should be tackled separately. In the language of OML's, apart from the finite-lattice trivial case, this issue should be translated to the so-called block-finite lattices. Consequently, it would be an interesting and challenging task to find indications for this case also deriving from the quantum-mechanical operator algebras perspective.

Having formulated all of these in the language of category theory, one obtains an opportunity to use it in its full power and explore the functoriality between **PBool** and n-**Mfd**. Particularly, it would be interesting to study smooth invariants of general four-dimensional manifolds from this perspective. We already referred to the fact that automorphisms (more generally, homomorphisms) of Boolean subalgebras should correspond to diffeomorphisms of charts of X; globally, automorphisms of an OML $\mathbb{L}$ would correspond to diffeomorphisms of whole X. This is potentially a very fertile area of research, since set- and model-theoretical tools seem to be inevitable in the study of the objects related $\mathrm{Aut}(\mathbb{L})$ (in the last two decades one could observe a real explosion of new results concerning the applications of sophisticated set-theoretical methods to automorphisms of a Calkin C^*-algebra and, at the same time, of a Boolean algebra $\mathscr{P}(\omega)/\mathrm{fin}$ (for a thorough discussion see [18]). The task of connecting these to the study of smooth manifolds invariants seems to be a very promising area of research.

Acknowledgements The authors are indebted to Torsten Asselmeyer-Maluga for very important questions regarding exotic smoothness that improved the text.

References

1. T. Asselmeyer-Maluga, Smooth quantum gravity: exotic smoothness and quantum gravity, in *At the Frontier of Spacetime*, ed. by T. Asselmeyer-Maluga (Springer, Cham, Switzerland, 2016), pp. 247–308
2. T. Asselmeyer-Maluga, C.H. Brans, Cosmological anomalies and exotic smoothness structures. Gen. Rel. Grav. **34**, 1767–1771 (2002)
3. T. Asselmeyer-Maluga, C.H. Brans, *Exotic Smoothness and Physics* (World Scientific Singapore, Singapore, 2007)
4. T. Asselmeyer-Maluga, J. Król, On the origin of inflation by using exotic smoothness. https://arxiv.org/abs/1301.3628
5. T. Asselmeyer-Maluga, J. Król, Inflation and topological phase transition driven by exotic smoothness. Adv. High Energy Phys. **2014**, 1–14 (2014)
6. T. Asselmeyer-Maluga, J. Król, How to obtain a cosmological constant from small exotic $\mathbb{R}^4$. Phys. Dark Universe **19**, 66–77 (2018)
7. T. Asselmeyer-Maluga, J. Król, A topological approach to Neutrino masses by using exotic smoothness. Mod. Phys. Let. A **34**(13), 1950097 (2019)
8. J.L. Bell, *Set Theory: Boolean-Valued Models and Independence Proofs* (Clarendon Press, Oxford, 2005)
9. K. Bielas, J. Król, A category-theoretic view on the smooth manifold coordinatization by Boolean subalgebras of an orthomodular lattice of projections (in preparation)
10. G. Birkhoff, J. von Neumann, The Logic of quantum mechanics. Ann. Math. **37**, 823–843 (1936)
11. P. Bongaarts, *Quantum Theory—A Mathematical Approach.* (Springer International Publishing Switzerland 2015)
12. C.H. Brans, Localized exotic smoothness. Class. Quant. Grav. **11**, 1785–1792 (1994)
13. M.L. Dalla Chiara, R. Giuntini, R. Greechie, *Reasoning in Quantum Theory: Sharp and Unsharp Quantum Logics* (Springer, Netherlands, 2004)
14. M. Davis, A relativity principle in quantum mechanics. Int. J. Theor. Phys. **16**, 867874 (1977)

15. S. De Michelis, M. Freedman, Uncountably many exotic $\mathbf{R}^4$'s in standard 4-space. J. Diff. Geom. **35**, 219–254 (1992)
16. A. Döring, J. Harding, Abelian subalgebras and the Jordan structure of a von Neumann algebra. https://arxiv.org/abs/1009.4945
17. G. Etesi, Exotica and the status of the strong cosmic censor conjecture in four dimensions. Class. Quantum Grav. **34**(24), 245010-1-245010-26 (2017)
18. I. Farah, E. Wofsey, Set theory and operator algebras, in *Appalachian Set Theory*, ed. by J. Cummings, E. Schimmerling (Cambridge University Press, 2012), pp. 63–120
19. M.H. Freedman, L.R. Taylor, A universal smoothing of four- space. J. Diff. Geom. **24**(1), 6978 (1986)
20. J. Harding, M. Navara, Subalgebras of orthomodular lattices. Order **28**, 549563 (2011)
21. C. Heunen, N.P. Landsman, B. Spitters, S. Wolters, The Gelfand spectrum of a noncommutative C^*-algebra: a topos-theoretic approach. J. Aust. Math. Soc. **90**, 39–52 (2011)
22. C. Heunen, N.P. Landsman, Spitters, B: a topos for Algebraic Quantum Theory. Commun. Math. Phys. **291**, 63–110 (2009)
23. G. Kalmbach, *Orthomodular Lattices* (Academic Press, London, 1983)
24. V. Karakostas, E. Zafiris, Contextual semantics in quantum mechanics from a categorical point of view. Synthese **194**, 847886 (2017)
25. J. Król, T. Asselmeyer-Maluga, K. Bielas, From quantum to cosmological regime. The role of forcing and exotic 4-smoothness. Universe **3**, 31 (2017)
26. A. Kupers, Lectures on diffeomorphism groups of manifolds. http://www.math.harvard.edu/~kupers/teaching/272x/book.pdf
27. T. Leinster, *Higher Operads Higher Categories* (Cambridge University Press, Cambridge, 2004)
28. T. Leinster, *Basic Category Theory* (Cambridge University Press, Cambridge, 2014)
29. A.J. Lindenhovius, Classifying finite-dimensional C*-algebras by posets of their commutative C*-subalgebras. Int. J. Theor. Phys. **54**, 46154635 (2015)
30. M. Nakahara, *Geometry Topology and Physics* (IoP Publishing London, 2003)
31. M. Pavičić, Bibliography on quantum logics and related structures. Int. J. Theor. Phys. **31**, 373455 (1992)
32. H. Primas, *Chemistry, chanics and Reductionism—Perspectives in Theoretical Chemistry* (Quantum MSpringer-Verlag, Berlin, Heidelberg, 1983)
33. C.H. Randall, D.J. Foulis, Tensor products of quantum logics do not exist. Notices of the American Math. Soc. **26** (1979)
34. J. Sładkowski, Gravity on exotic $\mathbb{R}^4$ with few symmetries. Int. J. Mod. Phys. D **10**, 311–313 (2001)
35. Svozil, K, *Quantum Logic*. (Springer, Singapore, 1998)
36. C.H. Taubes, Gauge theory on asymptotically periodic 4-manifolds. J. Diff. Geom. **25**, 363–430 (1987)
37. G. Takeuti, *Two Applications of Logic to Mathematics* (Princeton University Press, Princeton, 1978)
38. B. van den Berg, C. Heunen, Noncommutativity as a colimit. Appl. Categor. Struct. **20**, 393–414 (2012)
39. E. Zafiris, Boolean coverings of quantum observable structure: a setting for an abstract differential geometric mechanism. J. Geom. Phys. **50**, 99–114 (2004)

Chapter 8
Beyond the Space-Time Boundary

Michael Heller and Jerzy Król

Abstract In General Relativity, a space-time M is regarded as singular if there is an obstacle that prevents an incomplete curve in M from being continued. Such a space-time is completed to form $\bar{M} = M \cup \partial M$ where ∂M is a singular boundary of M. The standard geometric tools on M do not allow one "to cross the boundary". However, the so-called Synthetic Differential Geometry (SDG), a categorical version of standard differential geometry based on intuitionistic logic, has at its disposal tools permitting this to be done. Owing to the existence of infinitesimals, one is able to penetrate "germs of manifolds" that are not visible from the standard perspective. We present a simple model showing what happens "beyond the boundary" and when the singularity is finally attained. The model is purely mathematical and is mathematically rigorous but it does not pretend, at its present stage, to refer to the physical universe.

8.1 Introduction

There is a general agreement among specialists that "to say that a space-time is singular means that there is some positive obstacle that prevents an incomplete curve continuing" [2].[1] Very roughly speaking, there are two kinds of such obstacles: (1) some magnitudes, such as curvature or some scalars constructed from it, become unbounded along a timelike curve before it ends, or (2) a pathological behaviour of the differential structure of space-time prevents a timelike curve from being

[1]The present paper is based on a talk delivered at the conference "Category Theory in Physics, Mathematics and Philosophy" held at the Warsaw University of Technology, 16–17 November 2017.

M. Heller (✉)
Copernicus Center for Interdisciplinary Studies, ul. Szczepańska 1/5, 31-011 Cracow, Poland
e-mail: mheller@wsd.tarnow.pl

J. Król
University of Information Technology and Management, ul. Sucharskiego 2, 35–225 Rzeszów, Poland
e-mail: iriking@wp.pl

M. Kuś and B. Skowron (eds.), *Category Theory in Physics, Mathematics, and Philosophy*, Springer Proceedings in Physics 235,
https://doi.org/10.1007/978-3-030-30896-4_8

prolonged. The book by Clarke [2] is almost exclusively devoted to making precise and understanding the second of these obstructions. In the present paper, we continue this line of research but we essentially change the method of investigation. In the meantime (after the publication of Clarke's book) a new approach in mathematics has matured that not so much solves but rather circumvents many of the problems related to differentiability. We have in mind the so-called Synthetic Differential Geometry (SDG), an "extension" of the usual differential geometry, based on category theory (for which the fundamental monographs are [5, 6, 8]). The fact that this approach enforces the employment of intuitionistic logic results in the enriching of the real line $\mathbb{R}$ with various kinds of infinitesimals. We may imagine that they constitute the entire world inside every point of $\mathbb{R}$, a sort of a fiber over $x \in \mathbb{R}$. Owing to the existence of infinitesimals, differentiation becomes a purely algebraic operation and every function is differentiable as many times as required. This creates a unique opportunity to tackle the problem of space-time prolongations[2] and singularities in General Relativity (GR).

We approach this problem along the following lines. In Sect. 8.2, we give necessary preliminaries: we define these kinds of infinitesimals that are used in the sequel, we define the "kth order neighbouring relation", and the concept of the monad in terms of it—a kind of minimal portion of space. We also quote some of its properties.

In Sect. 8.3, we present an "infinitesimal version" of the usual differential manifold concept, called the formal n-dimensional manifold. In such a formal manifold, monads are domains of local maps.

In Sect. 8.4, we present a model which shows, by incorporating the machinery sketched in the previous sections, what happens "beyond" the singular boundary ∂M of space-time M. We assume that this boundary contains a strong curvature singularity such as the Big Bang type of singularity.

We add two appendices. The first appendix gives a more detailed mathematical description of how the transition from M to ∂M could look like. The second appendix presents a concept that could be useful in studying the dynamics of monads.

Although our model attempts to imitate, for pedagogical reasons, the standard evolution of the Friedman–Lemaître–Robertson–Walker (FLRW) cosmological model, it does not pretend to describe the actual evolution of the universe. At its present stage of development it is nothing more than a toy model.

[2]Usually, in differential geometry, one *extends* a semi-Riemannian manifold so that certain incomplete curves are given complete extensions. Here we *prolong* a space-time manifold by putting it, with the help of a suitable functor, in a category which is a model for SDG.

8.2 Preliminaries

In SDG, one considers various kinds of infinitesimals.[3] Let us denote by R the real line $\mathbb{R}$ enriched by infinitesimals.[4] In the present study, we focus on the following ones

$$D = \{x \in R | x^2 = 0\},$$

$$D_k = \{x \in R | x^{k+1} = 0\}, \ k = 1, 2, 3, ...,$$

$$D(n) = \{(x_1, ..., x_n) \in R^n | x_i x_j = 0, \ \forall i, j = 1, 2, 3, ..., n\},$$

$$D_k(n) = \{(x_1, ..., x_n) \in R^n | \text{ the product of any } k+1 \text{ of } x_i \text{ is } 0\},$$

and finally,

$$(D_\infty)^n = \bigcup_{k=1}^{\infty} D_k(n).$$

We can also define $D(V)$ and $D_k(V)$ for any finite dimensional vector space V (which is an R-module isomorphic to some R^n) [6, p. 15].

We can imagine infinitesimals as internal "degrees of freedom" of a single point of $\mathbb{R}$.[5]

In what follows, our important tool is the "kth order neighbouring relation", defined as

$$u \sim_k v \Leftrightarrow u - v \in D_k(V),$$

for $u, v \in V$. This relation is reflexive and symmetric, but it is not transitive; instead we have

$$(u \sim_k v \wedge v \sim_l w) \Rightarrow (u \sim_{k+l} w).$$

We assume that everything in this section happens in a "suitable" category $\mathscr{E}$. "Suitable" means a category equipped, among others, with a commutative ring object R, usually a topos. We further assume that the category $\mathbb{M}$ of manifolds (and smooth functions) can be regarded as a subcategory of $\mathscr{E}$. Let M be an n-dimensional (formal) manifold considered as an object of $\mathscr{E}$. We are interested in the "smallest neighbourhoods" in M. A good tool to investigate such neighbourhoods is the neighbourhood relation $\sim_k$ but it should first be generalized to the manifold context. Let $x, y \in M$ and k be a non-negative natural number; the relation $x \sim_k y$ holds in M iff there

[3] In general, infinitesimal objects are identified with the spectra of Weil algebras; see Appendix 1.

[4] R is a commutative ring but not a field. From the perspective of the category of sets, R is $\mathbb{R} \oplus \epsilon\mathbb{R} \oplus ...$, with ϵ the dual number, which is a Weil algebra. In the axiomatic presentation of SDG, R is a ring which in some categories, serving as models of a given axiomatic system, can be local or not.

[5] In the category of sets, R is represented as $\mathbb{R} \oplus \epsilon\mathbb{R} \oplus ...$ and this has a structure of a bundle with base $\mathbb{R}$.

exists a coordinate chart $f : U \to M$ such that $U \subseteq R^n$ is open[6] and in U one has $x' \sim_k y'$ with $f(x') = x$ and $f(y') = y$. If $k = 1$, we simply write $x \sim y$.

Let M be a manifold and $x, y, z \in M$. The neighbouring relations $\sim_k$ satisfy the following conditions

1. $x \sim_0 y$ iff $x = y$ (reflexivity).
2. $x \sim_k y$ implies $y \sim_l x$ if $k \leq l$ (symmetry),
3. $x \sim_k y$ and $y \sim_l z$ implies $x \sim_{k+l} z$ (quasi-triangle formula).

Since these conditions are akin to the usual concept of distance, we can define a "quasi-distance" function in the following way

$$\text{dist}(x, y) \leq k \ \text{ iff } \ x \sim_k y.$$

Since $D_k(n) \subseteq D_l(n)$ if $k \leq l$, the function "dist" determines a "size" of an object $D_k(n)$. Let us also notice that this quasi-metric is "quantised" (discrete) since it has its values in $\mathbb{N}$.[7]

Now, we define a few key concepts for our further considerations.

Let $x \in M$. The k-monad around x is defined to be

$$\mathcal{M}_k(x) := \{y \in M | x \sim_k y\} \subseteq M.$$

If $k = 1$, we write $\mathcal{M}(x)$. We also define $\mathcal{M}_\infty = \cup_{i=1}^{\infty} \mathcal{M}_i$ and assume that $x \sim_\infty y$ makes sense. We obviously have $y \in \mathcal{M}_k(x) \Leftrightarrow x \in \mathcal{M}_k(y)$. If $f : M \to N$ is a map between manifolds M and N then $x \sim_k y$ implies $f(x) \sim_k f(y)$, since in SDG every map $f : D_k \to R$ such that $0 \mapsto 0$, factorizes through D_k [5, Cor. 6.2].

The "kth neighbourhood of the diagonal", $M_{(k)} \subseteq M \times M$, is defined to be

$$M_{(k)} := \{(x, y) \in M \times M | x \sim_k y\}.$$

If V is an n-dimensional vector space, there is a canonical isomorphism

$$M_{(k)} \cong M \times D_k(V)$$

given by

$$(x, y) \mapsto (x, y - x),$$

and consequently there is an isomorphism between $\mathcal{M}_k(x)$ and $D_k(V)$ [6, p. 39]. Let us also notice that the quasi-distance $\text{dist}(d_1, d_2) \leq k$ introduces a partial order in $\mathcal{M}_\infty(x)$ (by inclusions).

[6]If V is a finite dimensional vector space, $U \subseteq V$ is formally open if $x \in U$ and $y \sim_k x$ implies $y \in U$. In general, the problem of topology in SDG is a tricky problem (see [1], however, for our case it is enough to consult [6]). Anyway, in a category like $\mathcal{E}$ topology is well defined (see [8, p. 120]).

[7]"...this 'dist'-function is almost like an $\mathbb{N}$-valued metric" [6, p. 83].

8.3 Manifold down to the Smallest Scale

The existence of infinitesimals essentially enriches the structure of differential manifolds. It enables the following definition ([4], see also [6, pp. 68–71]). An object M in the category $\mathcal{E}$ is said to be a k-formal n-dimensional manifold if, for each $x \in M$, there exists a monad $\mathcal{M}_k(x)$, isomorphic to $D_k(n)$, and a map $f : \mathcal{M}_k(x) \to M$. A bijective map $D_k(n) \to \mathcal{M}_k(x)$ onto a monad around x, mapping 0 to x, is said to be a k-frame at x. k can assume the value ∞. If $\mathcal{M}_\infty(x)$, we speak of a formal n-dimensional manifold (without specifying k).

It can easily be seen that R^n, for every n, is a formal n-dimensional manifold, and the monad $\mathcal{M}_\infty(v)$ around $v \in R^n$ is $\mathcal{M}_\infty(v) = v + D_\infty(n)$.

We have the following natural, but important, results.

If M and N are formal manifolds of dimensions m and n, respectively, then $M \times N$ is an $(m+n)$-dimensional formal manifold; and the monad around $(x, y) \in M \times N$ is $\mathcal{M}(x) \times \mathcal{M}(y)$ which is isomorphic to $D_\infty(m+n)$.

If M is a formal n-dimensional manifold, then its tangent bundle M^D is a $(2n)$-dimensional formal manifold [4, Proposition 1.4].

8.4 A Model

Let us consider a singular space-time; it is singular in the sense that it contains at least one incomplete (timelike or null geodesic) curve that cannot be continued in any extension of this space-time. The regular (non-singular) part of this space-time forms a differential manifold M. We define the completion of M as $\bar{M} = M \cup \partial M$ and call ∂M the singular boundary of M. We assume that this boundary is attainable from M, i.e. that M is open and dense in $\bar{M}$. Details of this construction are of no importance for our further analysis (several proposals are known, such as g-boundary, b-boundary, causal boundary, and others [2, 3]).

A singular boundary can contain, besides endpoints of inextendible curves that cannot be continued in any extension of space-time, also endpoints of curves that can be continued in some of its extensions, and "points at infinity". In what follows, we assume, for simplicity, that ∂M contains only the endpoints of inextendible curves that cannot be continued in any extension of space-time.

For the sake of concreteness, let us think about the FLRW space-time with the Big Bang singularity (strong curvature singularity) in the beginning (the central Schwarzschild singularity would also fit the picture), and let us contemplate the evolution of the universe back in time. Everything happens according to the standard cosmological model. The universe shrinks, subsequent cosmic eras succeed each other. Finally, the contraction attains the state in which differential properties of space time break down completely. The universe leaves the "manifold region" M and enters its boundary ∂M. This means that the standard smooth manifold description breaks down, and we assume that at this stage the category $\mathcal{E}$ takes over (see Appendix 1). The contraction has reached such a degree that infinitesimals enter into play. We can thus employ methods of SDG to gain an insight into what is going on. The general picture that emerges is the following.

After crossing ∂M, domains U of local charts (already in the category $\mathscr{E}$) become infinitesimal, and the manifold becomes a 4-dimensional formal manifold as defined in Sect. 8.3. Local charts are now of the form $\mathscr{M}_\infty(x) \to M$ (the fact that we use the same letter for denoting the space-time manifold and the formal manifold should not lead to misunderstandings). But the universe continues shrinking, and finally its size reduces to a single monad $\mathscr{M}_\infty(x_0)$.[8] The contraction goes on, but now only in the sense of the metric $\text{dist}(d_1, d_2) \leq k,\ d_1, d_2 \in \mathscr{M}_\infty(x_0)$. This means that the differentiability properties are lower and lower, and we obtain a decreasing sequence

$$\mathscr{M}_\infty(x_0), \ldots, \mathscr{M}_k(x_0), \mathscr{M}_{k-1}(x_0), \mathscr{M}_{k-2}(x_0), \ldots.$$

Finally, when the contraction produces $\mathscr{M}_0(x_0)$, the process comes to a halt, since $\mathscr{M}_0(x_0) = \{y \in R | x_0 \sim_0 y\}$ which, in turn gives $x_0 = y$, and all quasi-distances $\text{dist}(x_0, y)$ reduce to zero.

It is instructive to follow the entire process starting from the zero-state. It seems natural to regard the increasing sequence of ks as a sort of quantised time. If this seems too farfetched, one can treat "time" as just the name of a parameter. However, it is important to notice that k, in fact, means the "degree of differentiability", the place at which the Taylor expansion truncates (all higher order terms vanish). Each subsequent instant of this time improves the differential properties of the process.

It is astonishing that the transition from $k = 0$ to $k = 1$ is so exuberantly reach. Having at our disposal the very first degree of differentiability, we can do large parts of affine geometry, affine connection included, combinatorial differential forms, tangent bundle and lot of differential geometry (in fact, a substantial part of Kock's seminal SDG monograph [6] is limited to exploring the geometry of the first order neighbourhoods).

Of course, when we jump to $k = 2$, differentially-geometric properties substantially improve. Some aspects of metric geometry come into force [6, Chap. 8].

For doing k-jet theory, we evidently need a sufficiently high k.

Finally, when $k \to \infty$, we end up in $\mathscr{M}_\infty(x_0)$ and we have differentiability of infinite degree. As the universe expands, it goes through the phase of a formal manifold, and when infinitesimals cease to play any role (because of the expansion), the standard smooth manifold regime takes over. In the language of the space-time boundary, this means that the universe goes from ∂M to M. It is here that we should place the transition from the category $\mathscr{E}$ to the category SET, and possibly identify this with what physicists call Planck's threshold.

Acknowledgements The authors express their thanks to Professor Samuel Fletcher for his penetrating remarks which helped to improve the paper.

---

[8] Perhaps an explanation is here helpful. A formal manifold can be defined as an object M for which there exists a family of jointly surjective maps $\{f_i : U_i \to M\}_{i \in I}$ [5, p. 68]. The "shrinking of the size" means that we formally reduce the family to a single map $f_p : U_p \to M,\ p \in I$, i.e. the formal manifold is reduced to U_p, but U_p is now $\mathscr{M}_\infty(x_0)$ where x_0 is a global point at the boundary (the unique global point at the monad).

8.1 Appendix: Through the Boundary

In this appendix, we give a short mathematical description of what the transition from M to ∂M could look like. The crucial point is that if we go from space-time M to its singular boundary ∂M, we must switch from the category SET of sets and maps between sets as morphisms to a suitable category, to which we have assigned the symbol $\mathcal{E}$, the internal logic of which is intuitionistic what enables infinitesimals to appear.

Space-time M is supposed, as always, to be a smooth paracompact manifold. First, we move from its description in terms of maps and atlases to the functional description in terms of the algebra $C^\infty(M)$ of smooth functions on M. This is possible owing to the generalized Gelfand-Naimark theorem which asserts that the category of locally compact Hausdorff spaces and proper continuous maps is anti-equivalent to the category of commutative C^*-algebras and nondegenerate morphisms [7]. We recall that map $\phi : X \to Y$ between locally compact Hausdorff spaces is said to be proper if, for any compact $K \subset Y$, $\phi^{-1}(K)$ is compact in X. And a morphism $\psi : A \to B$ between C^*-algebras is said to be nondegenerate, if the linear span of all expressions of the form $\psi(a)b$, $a \in A$, $b \in B$, is dense in B. Two categories G and H are said to be anti-equivalent (or dual) if there exist contravariant functors $\alpha : G \to H$ and $\beta : H \to G$ such that $\alpha \circ \beta$ and $\beta \circ \alpha$ are naturally isomorphic to id_H and id_G, respectively.

We now look at $C^\infty(M)$ from another perspective. A smooth algebra (or C^∞-ring) is an algebra A over $\mathbb{R}$ for which the product $\cdot : \mathbb{R} \times \mathbb{R} \to \mathbb{R}$ lifts to the algebra product $A \times A \to A$, and also every smooth map $f : \mathbb{R}^n \times \mathbb{R}^m$ lifts to a map $A(f) : A^n \to A^m$, in such a way that projections, identities and compositions are preserved. Formally, such a smooth algebra is a functor from the category Cart of Cartesian spaces to the category SET,

$$A : \text{Cart} \to \text{SET},$$

that preserves finite products. The category of such functors as objects and natural transformations between them as morphisms is denoted by C^∞-Alg [8, pp. 15–16].

If $M \in \mathbb{M}$ is a smooth manifold then the functor

$$C^\infty(M) = \text{Hom}_{\mathbb{M}}(M, -)$$

is a smooth algebra $C^\infty(M)$. Considering an object (here a smooth manifold) in another category may not be a superficial change. Changing a categorical context can not only provide new tools of investigation, but can also affect the properties of the object itself. Let us pursue this line of research.

A smooth algebra A is said to be finitely generated if it is of the form $C^\infty(\mathbb{R}^n)/I$, for $n \in \mathbb{N}$ and an ideal I; if additionally I is finitely generated, A is said to be finitely presented. For every smooth manifold M, the smooth algebra $C^\infty(M)$ is finitely presented [8, p. 25].

Finitely generated smooth algebras and C^∞-homomorphisms between them as morphisms form a category denoted by FGAlg (see [8, p. 21]). We define the category LOC (of loci) as the opposite category with respect to FGAlg. The objects of LOC are the same as those of FGAlg (if $A \in$ FGAlg, we shall write lA when A is considered as an object of LOC), but morphisms are reversed. Since the functor from the manifold category $\mathbb{M}$ to C^∞-Alg, given by $M \mapsto C^\infty(M)$, is contravariant, to obtain the correct variance, we switch to the category LOC. And indeed, we have the covariant functor $s : \mathbb{M} \to$LOC, given by $M \mapsto lC^\infty(M)$ which, importantly, is full and faithful [8, p. 60].

This switching of the categorical environment produces radical changes. A new element that appears in this context is the so-called Weil algebra. It is a finite dimensional $\mathbb{R}$-algebra W having a maximal ideal I such that $W/I \simeq \mathbb{R}$ with $I^n = 0$ for some $n \in \mathbb{N}$. It has a unique smooth algebra structure and is finitely presented. Objects of LOC (smooth loci) corresponding to Weil algebras are infinitesimal spaces. In this way, infinitesimals appear in our model. We immediately have: $C^\infty(\mathbb{R}) \in$ LOC is a real line enriched with infinitesimals which we denote by R.

However, $s(M)$, the image of M under the functor s, is much richer than M. It contains infinitesimal portions of a manifold or spaces which can be called "germs of a manifold". To define them, let us notice that if $lA = C^\infty(\mathbb{R}^n)/I$ then $p : 1 \to lC^\infty(\mathbb{R}^n)$ is a point of lA, where $1 = lC^\infty(\mathbb{R}^0)$. Let now $p \in \mathbb{R}$. We define the germ $C^\infty_p(\mathbb{R})$ of R at p to be $\bigcap\{s(U)|p \in U \text{ open in } \mathbb{R}\}$. And analogously for $p \in \mathbb{R}^n$. We are now ready to define the germ of lA at $p \in \mathbb{R}^n$ as $lC^\infty_p(\mathbb{R}^n) \cap lA$. This of course remains valid if $lA = s(M)$.

Let us consider a function $f : M \to \mathbb{R}$, and the germ f_p of this function at $p \in M$ (in the usual sense). This germ can now be identified with the restriction of the function f to the germ of the manifold M at p (for details of which see [8, p. 64]).

Usually, to improve the geometric properties of a given model,[9] one once again changes the category LOC to the category $\mathrm{SET}^{\mathrm{LOC}^{op}}$ of presheaves on LOC or to some of its subcategories [8]. In our case, this does not seem indispensable since LOC has good properties if limited to "sufficiently small" spaces [8, p. 71].

8.2 Appendix: Non-holonomous Monads

In studying interactions between different levels of differentiability in a monad, the following concept is useful.

Let M be a (formal) manifold, and let us consider a sequence $k_1, \ldots, k_r$ of non-negative integers. Let us also consider the set $M_{(k_1,\ldots,k_r)} \subseteq M^{r+1}$, the elements of which are $r+1$-tuples

$$X = (x_0, k_1, \ldots, k_r)$$

[9] For instance, to have a Cartesian closed category with power objects.

such that

$$x_0 \sim_{k_1} x_1 \sim_{k_2} \cdots \sim_r x.$$

If $r = 1$, we recover M_{k_1}. Assigning to such an $r + 1$-tuple its first element x_0 (when x_0 varies over M), we obtain a bundle $M_{(k_1,\dots,k_r)} \to M$ over M. The fiber over $x = x_0$ of this bundle is called the non-holonomous $(k_1, \dots, k_r)$-monad around x and is denoted by $\mathscr{M}_{k_1,\dots,k_r}(x)$. There exists a map

$$\mathscr{M}_{(k_1,\dots,k_r)}(x) \to \mathscr{M}_{(k_1+\dots+k_r)}$$

transforming non-holonomous monads into ordinary (holonomous) ones (for more see [6, pp. 86–88]).

References

1. M. Bunge, F. Gago, *Synthetic Differential Topology* (Cambridge University Press, Cambridge, 2018)
2. C.J.S. Clarke, *The Analysis of Space-Time Singularities* (Cambridge University Press, Cambridge, 1993)
3. C.T.J. Dodson, Space-time edge geometry. Int. J. Theor. Phys. **17**(6), 389–504 (1978)
4. A. Kock, Formal manifolds and synthetic theory of jet bundles. Cahiers de topologie et géometrie différentielle catégoriques **21**(3), 227–246 (1980)
5. A. Kock, *Synthetic Differential Geometry*, 2nd edn. (Cambridge University Press, Cambridge, 2006)
6. A. Kock, *Synthetic Geometry of Manifolds* (Cambridge University Press, Cambridge, 2009)
7. N.P. Landsman, *Lecture Notes on C*-Algebras and K-Theory*, http://www.math.nus.edu.sg/~matwml/Research_Materials/Operator
8. I. Moerdijk, G.E. Reyes, *Models for Smooth Infinitesimal Analysis* (Springer, New York, 2010)

Chapter 9
Aspects of Perturbative Quantum Gravity on Synthetic Spacetimes

Jerzy Król

Abstract Synthetic differential geometry allows for infinitesimal monads in the usual structure of Minkowski spacetime M^4. Gravity, as a field defined on M^4, can explore monads. Moreover, at high energies gravity shows its quantum character and interacts via interchanging gravitons on monads. Such defined interactions lead to important consequences. Pure perturbative quantum gravity (QG) on synthetic spacetimes is finite. Certain supersymmetric perturbative QG's with matter are also finite on the synthetic spacetime with monads. The case of non-supersymmetric perturbative QG with matter is not solved decisively here. However, at least for the two cases above, classical gravitational interactions are defined via quantization of the theories which solves the conundrum of incompatibility of Einstein equations with quantum matter sources.

9.1 Introduction

Einstein equations (EE) with a cosmological constant Λ

$$R_{\mu\nu} - \frac{1}{2}Rg_{\mu\nu} + \Lambda g_{\mu\nu} = \frac{8\pi G}{c^4}T_{\mu\nu}$$

teach us that geometry of spacetime, i.e. the metric $g_{\mu\nu}$ at the left hand side, is determined by the distribution of matter and energy at the right hand side. The energy-momentum tensor $T_{\mu\nu}$ is, however, determined by quantum fields in spacetime while $g_{\mu\nu}$ as well as the Ricci tensor $R_{\mu\nu}$ and the scalar tensor R are purely classical objects. Attempts to quantize general relativity (GR) as a 4-dimensional quantum field theory (QFT) on Minkowski spacetime have failed. Thus we would be left with the

J. Król (✉)
University of Information Technology and Management, ul. Sucharskiego 2, 35-225 Rzeszów, Poland
e-mail: iriking@wp.pl

M. Kuś and B. Skowron (eds.), *Category Theory in Physics, Mathematics, and Philosophy*, Springer Proceedings in Physics 235,
https://doi.org/10.1007/978-3-030-30896-4_9

uncomfortable situation of not knowing the quantum shape of the gravitational field while all other fields of matter, appearing in EE via $T_{\mu\nu}$, are quantum fields obtained by well-defined quantization of their classical counterparts. This is a kind of inconsistency of the equations facing the quantum realm of the world. It should be cured somehow if one thinks seriously about a consistent theory unifying gravity with other forces. One reason allowing for considering gravity as purely classical, even at regime where other interactions are quantum, is the incredible weakness of the gravitational interactions compared to the others. In such case we would be left with classical spacetime defined by EE and various fields which undergo routine quantization procedures known from QFT. One approach to quantizing gravity is taking the background spacetime, which can be chosen to be Minkowski 4-dimensional smooth manifold M^4, and let the metric fluctuate. Thus the flat metric $\eta_{\mu\nu} = \mathrm{diag}(-1, +1, +1, +1)$ allows for the fluctuations $h_{\mu\nu}$ such that a general metric reads

$$g_{\mu\nu} = \eta_{\mu\nu} + h_{\mu\nu} .$$

Then one tries to quantize $h_{\mu\nu}$ as any other field on M^4. The quantization should lead eventually to a *perturbative quantum gravity in dimension 4*—the miracle object which, unfortunately, does not seem to exist. The main problem is that Feynman diagrams of this theory, which are main tool of any perturbative QFT, are generically divergent and the divergences can not be cured by any known process of renormalization. One reason for that is the nonlinear interaction of gravity with itself and with any other form of energy. Another problem is the dimensional nature of the gravitational coupling constant, i.e. the Newton constant G. More precise description of that behavior is as follows (e.g. [17])

- Pure perturbative QG, i.e. $T_{\mu\nu} = 0$ and the only quantized field is $h_{\mu\nu}$, is finite at one loop. It is divergent at two and more loops and it is a nonrenormalizable theory.
- Perturbative QG with matter fields ($T_{\mu\nu} \neq 0$) is divergent already at one loop and it is a nonrenormalizable theory.

Thus one meets two seemingly unavoidable obstacles when trying to describe gravity as a fundamental theory. One is the inherent nonrenormalizability of pure or matter perturbative QG, i.e. quantized GR in dimension 4, while the second is inconsistency of EE where gravity remains classical and the other fields are quantized. Additionally, and connected with the above, one does not have clear insights on what nonperturbative 4-dimensional QG would be like.[1] In this chapter we address all three above issues but with quite unusual technique based on synthetic differential geometry (SDG).

[1]Loop QG does not seem to be a complete nonperturbative QG since it breaks full 4-dimensional diffeomorphism invariance.

9.2 Synthetic Differential Geometry, Spacetime and Gravity

Classical gravity is inherently and inseparably connected with the geometric structure of spacetime. That is why we expect that they strongly influence each other also at the quantum regime. Even though spacetime may not exist at all, or may be substantially modified as a smooth manifold at the Planck regime, the modification should follow from what happens with gravity at such regime. Here we present main ideas behind synthetic enrichment of the spacetime structure and its influence on the issue of existence of a theory of quantum gravity. Details will be presented in the forthcoming paper with Michael Heller.

9.2.1 SDG and the Perturbative QG

Any theory in mathematics is formulated by using certain formal language and corresponding logic. Almost all mathematics can be expressed in the language of set theory i.e. in the language that all of Zermelo–Fraenkel axioms of set theory are written in plus, optionally, the axiom of choice (AC). The logic is the *classical* first order predicate logic. From the point of view of category theory these two concepts are not independent but rather they create the category SET with its internal logic being the classical logic and the class of objects comprising all sets with morphisms being functions. This dependence of logic and sets can be further studied from the point of view of different categories than SET. In particular, weakening the logic by strict refraining from the use of AC is paralleled by the strict avoiding the excluded middle law, i.e. assuming that for any sentence p, the sentence $p \vee \neg p$ can be 'not true', i.e. $p \vee \neg p \leq 1$. Such logic without excluded middle law and set theory without AC is characteristic for *intuitionistic* reasoning or logic. This stays in sharp contrast to classical logic and set theory of SET. Categories where it is possible are known as *toposes*. The particularly well recognized and studied class of toposes are *Grothendieck toposes*. Their internal logic is again intuitionistic, but they also have canonically defined object of natural and real numbers as certain varying sheaves of sets (rather than constant sets in SET). Without incorporating the intuitionistic reasoning one can not enter the world in which $\mathbb{N}$ and $\mathbb{R}$ show impossible (from the classical point of view) properties. One crucial example below is the existence of intuitionistic real numbers $d \in R$ for which $d^k = 0$ with $d \neq 0, k \in \mathbb{N} \setminus \{0\}$. The 'collection' of such reals comprises the subobject of R (in special toposes) called k-monads. The suitably modified differential geometry in toposes allowing for monads is just *synthetic differential geometry* and the corresponding methods of such toposes are called *synthetic*. Main objective of this paper is to allow for such weakening of logic and set theory in physical theories and to study the consequences of such choice. The particularly well-defined area of applicability of synthetic methods will appear to be perturbative quantum gravity.

Let us be more precise. In classical mathematics the real line $\mathbb{R}$ is the model of the theory of complete ordered fields. Leaving aside the axiom of choice and the excluded middle law as above and thus working in intuitionistic environment we have plenty of categories—toposes where the internal mathematics becomes intuitionistic. In particular, in Grothendieck toposes there are objects of natural $\mathbb{N}$, and real numbers R. R is usually a ring which can be tailored as local and Archimedean (in some toposes) and contains infinitesimal subobjects $D_k \subset R$ of infinitesimal 'elements' [15]

$$D_k = \{d \in R : d^{k+1} = 0\},\ k = 0, 1, 2, \dots .$$

For any finite dimensional vector space $V \simeq R^n$ we can consider $D_k(V) = \{d = (d_1, \dots, d_n) \in R^n : d_i \in D_k, i = 1, 2, \dots, n\}$ which means $d \in D_k^n$. In what follows we assume that a topos $\mathscr{E}$ allows for developing internally all the machinery of SDG and considering R as the 'real line'. It follows that the products R^n parameterize locally the spacetime domains (in n dimensions), thus monads in such spacetime appear. These monads, in turn, are parameterized by D_k^n [12, 13]. More precisely, let M be a n-dimensional manifold in $\mathscr{E}$ and $x \in M$. The k-monad around x is defined to be

$$\mathscr{M}_k(x) := \{y \in M | x \sim_k y\} \subseteq M$$

where $x \sim_k y \iff x - y \in D_k^n$ so that there is an isomorphism between $\mathscr{M}_k(x)$ and $D_k(V)$ with D_k^n by sending $D_k(V) \ni 0 \mapsto x \in \mathscr{M}_k(x)$ [13, p. 39].

On the other hand, working entirely in SET, the objects D_k above can be represented as maximal ideals in the smooth quotient ring (Weil algebra) $C^\infty(\mathbb{R})/(x^{k+1})$ of the ring of smooth real functions on $\mathbb{R}$ by the ideal (x^{k+1}) generated by x^{k+1}, i.e. $x^{k+1} = 0$ (see the Appendix A and [15]). The object (ring) R of real numbers in $\mathscr{E}$ is represented in SET as $C^\infty(\mathbb{R})$. Using the notation for the generator of the ideals as ε (the dual number) we write

$$\begin{aligned} D_1 &= \varepsilon\mathbb{R} \\ D_2 &= \varepsilon\mathbb{R} \oplus \varepsilon^2\mathbb{R} \\ D_k &= \varepsilon\mathbb{R} \oplus \cdots \oplus \varepsilon^k\mathbb{R},\ k = 1, 2, \dots \end{aligned} \tag{9.1}$$

There exists a subtle interplay between the internal to $\mathscr{E}$ and external in SET presentations of the object D_k. Namely, given a physical theory on the spacetime M with monads $\mathscr{M}(x)$, $x \in M$ one faces important problems (in what follows we use the same letter M also for Minkowski spacetime). One is the issue of Poincaré invariance of the theory on M. Another, related with the first, is the question about physical fields propagating over monads and the physical effects of such propagation. However, there is one big meta-question: how is it possible that we enter the intuitionistic reasoning allowing for monads in a topos, thus leaving the usual SET environment and what is the impact of that leave on the smooth structure of spacetime. We analyze first two questions in the case of perturbative quantum gravity when Minkowski M^4

is replaced by the synthetic spacetime. The meta-question will be addressed in the forthcoming paper and we will give here only short commentaries.

Physical fields are defined as certain functions on spacetime. The linearization of gravity, i.e. $h_{\mu\nu}$ is also defined on M^4 hence we neglect GR at this stage and try to quantize $h_{\mu\nu}$ on flat M^4. This leads to the program of the covariant (Lorentz invariant) perturbative QG as firstly proposed and developed by Feynman [8, 9] (see also [7] and modern presentation [16]). There are known difficulties with renormalizability of such theory mentioned in the Introduction. Let us define such theory on synthetic spacetime with monads. To this end we consider gravitational linearized field $h_{\mu\nu}$ as capable of exploring the monads possibly at sufficiently high energies, i.e. of the order of Planck energy or higher. The main point is that $h_{\mu\nu}$ interacts with itself and with other fields on monads, which constrains the vertices of the corresponding Feynman diagrams to live entirely on monads. The decomposition $g_{\mu\nu} = \eta_{\mu\nu} + \sqrt{8\pi G} \cdot h_{\mu\nu}$, where $\sqrt{8\pi G}$ is the gravitational coupling constant, shows that in the units $\hbar = c = 1$ the Newton constant has units of $[m^2]$. The dual numbers ε on a monad (9.1) have natural units of the length $[m]$ and thus the Newton constant on a monad is represented as a quantity proportional to ε^2 having natural dimension of $[m^2]$. Let it be $G \cdot \varepsilon^2$. However, it follows that

$$(\sqrt{G})^{k+1} = 0 \;\text{ on }\; D_k,\; k = 1, 2, \dots .$$

The minimal choice is $k = 2$ since then $G^{3/2} = 0 \Longrightarrow G^2 = 0$, while on D_1, $\sqrt{G}^2 = 0 = G$ meaning that there are no gravitational interactions on D_2. Thus our first rule is that, at high energies, G lives on D_2 as the (quantity proportional to) squared dual number.

Let us apply this to the pure general relativity given by the Einstein Lagrangian (e.g. [7, p.86])

$$\mathscr{L} = \frac{1}{16\pi G^2} \int \sqrt{g} R d^4 x.$$

Under linearization of $g_{\mu\nu}$ and taking into account that any n-graviton, $n = 1, 2, \dots$, interaction vertex should appear in the Lagrangian, and using $f = \sqrt{8\pi G}$ for the coupling constant, we obtain the following expression [7]

$$\frac{1}{16\pi G^2} \sqrt{g} R[\varepsilon] = f^0 \cdot L_{(2)} + f \cdot L_{(3)} + f^2 \cdot L_{(4)} + \dots + f^{n-2} \cdot L_{(n)} + \dots . \tag{9.2}$$

By the rule sending G to $G \cdot \varepsilon^2$ we have $\sqrt{8\pi G} = f \to f \cdot \varepsilon$ on D_2, and we obtain the expansion on monads in the form

$$\frac{1}{16\pi G^2} \sqrt{g} R[\varepsilon] = (f \cdot \varepsilon)^0 L_{(2)} + (f \cdot \varepsilon) L_{(3)} + (f \cdot \varepsilon)^2 L_{(4)} =$$
$$L_{(2)} + (f \cdot \varepsilon) L_{(3)} + (f \cdot \varepsilon)^2 L_{(4)} \in \mathbb{R} \oplus \mathbb{R}[\varepsilon] \oplus \mathbb{R}[\varepsilon^2] = \mathbb{R} \oplus D_2.$$

which terminates at 4-vertex since $\varepsilon^k = 0,\ k \geq 3$.

The next step would be to build Feynman diagrams of this theory. Given 4-graviton vertex, it gives contribution $\sim f^2$ but trying to attach another 4-vertex in a tree diagram, it gives another factor $\sim f^2$ which together give rise to $\sim f^4$. When present on the same monad, it has to vanish. Similarly, any tree diagram containing 4-vertex and a 3-vertex has to vanish on a monad since its contribution $\sim f^3 = 0$. The non-vanishing contribution on a single monad comes from tree diagrams containing two 3-vertices with $\sim f^2 \neq 0$ in D_2.

9.2.2 Classical $g_{\mu\nu}$ as a Quantized Field

Let us analyze the outcome of the previous section in the linearized theory of gravity on monads from the point of view of Poincaré and gauge invariance. As follows from the Lagrangian (9.2) of pure QG, not only is there an infinite number of vertices, but also the simplest 3-vertex has around 90 terms [1]. This is a tremendous complication. On the other side, placing the pure QG on monads, as in the previous section, resulted in reducing the vertices into up to 3-vertices. Moreover, there should be maximally two 3-vertices at a single monad. Being a large simplification, it spoils the gauge invariance of the theory. The point is that when considering isolated Feynman diagrams (or isolated subset of them) their contributions usually are not gauge invariant. One possibility would be to allow for breaking Lorentz invariance, what is anyway a returning theme in the context of QG theories. However, we refuse this possibility and follow the fully 'covariant' QG as was also the intention of Feynman himself. A way to finding the solution is to refrain from using Feynman diagrams at all.

Let us recall that on a single monad there can be placed two 3-vertices (if there are more such 3-vertices this would give the vanishing contribution since $\varepsilon^{2+k} = 0$, $k \geq 1$). Other vertices are placed on other monads (still maximally two 3-vertices on a monad), so that the contributions from each monad are taken from $\mathscr{E}$ to SET by projection π (see (9.3) below). This means that the entire contribution from a bigger Feynman (tree) diagram is decomposed into separated monads contributions. Then the projection from a monad to SET relies on discarding ε^2 factors (or equivalently, taking $\varepsilon^2 = 1$). In other words, every many-3-vertices tree diagram can contribute in SET as without monads. Having clarified this point we are, however, still in a trouble since any Feynman diagram containing higher n-vertices vanishes spoiling the invariance of the theory at the tree-level.

The miracle happens when one turns to the complexified momenta. There emerge important extra-properties changing the entire approach:

i. all such complex momenta can be on-shell, i.e. there are no virtual gravitons;
ii. all tree amplitudes building the matrix $\mathbb{M}(z)$, $z \in \mathbb{C}$ (see Appendix B) are generated by 3-vertices diagrams;

iii. we do not need complicated totality of tree Feynman diagrams, rather there is the BCFW recursion formula gluing 3-vertices and determining all amplitudes in $\mathbb{M}(z)$ just because of the Poincaré invariance.

One could wonder whether it is possible at all to reformulate a quantum field theory such that everything happens on-shell and all amplitudes (tree level) are rather easily and efficiently calculated recursively. Some facts explaining such possibility are placed in the Appendix B. The BCFW formula has been derived in the recent years and the approach actively developed since then (see e.g. [1, 2, 6]).

Let us see how the BCFW algorithm helps solving problems with our formulation of synthetic pure QG. First, the initial data of the algorithm are 3-vertices. In Sect. 9.2 we saw that also in synthetic approach, when placing vertices on monads, there survive only a 3-vertex which can be glued with another 3-vertex on a single monad $\mathcal{M}(x)$. Any additional vertex is placed on a separate monad $\mathcal{M}(y)$, $x \neq y$. The momenta coordinates in SET live on the cotangent spaces $T^{\star}_{x,y}M$ to the Minkowski spacetime at points x, y which are isomorphic to the tangent spaces. In SDG the tangent space coordinates live on monads too (see [12]). To represent complex momenta on monads we need to complexify its structure. It is performed directly. The complexification of coordinates of a monad is just building the complex monadic structure $\mathcal{M}^{\star}(x)$ at x, parametrized by $D_2(4) \oplus i\, D_2(4)$. In terms of dual number generator ε it reads

$$\mathcal{M}^{\star}(x) = \{x\} \oplus \varepsilon\mathbb{R}^4 \oplus \varepsilon^2\mathbb{R}^4 \oplus \mathrm{i}\varepsilon\mathbb{R}^4 \oplus \mathrm{i}\varepsilon^2\mathbb{R}^4 .$$

Being equipped with complexified monads we can write down various Feynman diagrams in complex momenta on monads. But the BCFW formula (see (9.8) and (9.9) in Appendix B) shows much more. First, it gives the recipe how to combine the 3-vertices in complex momenta into arbitrary tree amplitudes. Second, in fact the bigger diagrams are not Feynman diagrams since we are using entirely on-shell complex momenta. In our monadic case, taking two such 3-vertices on a monad and using the SET projection π (see also (9.7) in Appendix A)

$$\begin{aligned} \pi : \mathscr{E} \to \mathrm{SET} \\ D_2 = \varepsilon\mathbb{R} \oplus \varepsilon^2\mathbb{R} \ni (\varepsilon \cdot a, \varepsilon^2 \cdot b) \mapsto (a, b) \in \mathbb{R}^2 \end{aligned} \tag{9.3}$$

one finds

$$\mathrm{BCFW}(v_1^{(3)} - - - - - - v_2^{(3)}) = \pi[\mathrm{BCFW}(v_1^{(3)} - - - - - - v_2^{(3)})]_{|\mathcal{M}(x)}$$

where $v_{1,2}^{(3)}$ are two 3-vertices (on monads and outside) composed according to the BCFW algorithm (on monads but also the outside). Repeating the validation of the BCFW procedure on every pair of 3-vertices on monads and applying the π projection we get finally literally the same result as for BCFW applied outside of the monads in complexified momentum space to every collection of 3-vertices. Moreover, as stated above, the main result of the BCFW schema is that the procedure covers *all* tree amplitudes of the pure QG [5]. That means the requirements of Poincaré invariance and that all processes to be on-shell, applied to the composed in pairs 3-vertices in the complexified momenta, leads to the entire tree-level of the theory. By the same argumentation we can conclude that pure perturbative QG on synthetic spacetime reproduces precisely the tree-level of the theory on ordinary Minkowski spacetime without referring to loop diagrams at all. This is basically the result of applying the synthetic methods to spacetime in the context of BCFW procedure. Our argumentation, in some sense, has come full circle. We started with classical GR with $g_{\mu\nu}$ where all effects of gravity can be (in principle) calculated from the Einstein equations. Then we quantized the linearized theory on synthetic spacetime taking the linearization $g_{\mu\nu} = \eta_{\mu\nu} + f \cdot h_{\mu\nu}$. As the result we have obtained the tree level of the quantum theory. But the tree level reproduces (in the limit of many-legs diagrams) classical effects of the theory described by $g_{\mu\nu}$ we started with. This tree level of QG is the complete answer the quantization procedure of GR on synthetic spacetime can give. The pure perturbative covariant QG on synthetic spacetime is entirely finite theory. However, the emerging classical level from quantum, improves fundamentally the status of Einstein equations. $g_{\mu\nu}$ can now be considered as fully quantized field and as such it would not conflict any longer with quantized matter fields in $T_{\mu\nu}$. However, we have dealt with pure QG where $T_{\mu\nu} = 0$ and the conflict has not been manifest so far. Thus we should analyze the remaining case of the matter perturbative QG where $T_{\mu\nu} \neq 0$.

9.3 Can Perturbative QG Be Any Fundamental Theory?

EE describe classical gravity. Main objection of this paper is that EE could describe as well the relation between quantized gravitational field and, presumably, the quantized matter fields giving contribution to $T_{\mu\nu}$. However, even in the case of pure QG the new fundamental role is assigned to $g_{\mu\nu}$. Namely, as we saw in the previous section $g_{\mu\nu}$ is at the same time a quantized field which leads to a new role of EE. This is the effect of synthetic methods applied to spacetime and linearized gravity. Can we extend the synthetic quantization of GR over theories with $T_{\mu\nu} \neq 0$?

In what follows we will see that this problem is not easily solved. Entire argumentation relies heavily on the BCFW rule. The rule, in turn, is derived from the analytical behavior of the S-matrix $\mathbb{M}(z)$ at infinity, namely

$$\lim_{z\to\infty} \mathbb{M}(z) = 0\,. \tag{9.4}$$

This vanishing at infinity guarantees that BFCW holds true (see the Appendix B). In pure QG (9.4) is true and BCFW is still valid. The same property (9.4) holds true also for $\mathcal{N} = 8$ supergravity with the presence of quantum fields giving $T_{\mu\nu} \neq 0$. That is why this supergravity theory formulated on synthetic spacetime is again finite and $g_{\mu\nu}$ quantized. For matter QG without supersymmetry, which is the most interesting case to us, we do not have any proof that (9.4) holds true. Actually it may not be true [1]. We should follow different way directly referring to the tree level of the theory. First, we should ensure ourselves that as long as this QG is formulated on synthetic spacetime, only the tree level of such covariant theory survives. However there appear complications. When one switches to the complex momenta as before then we are left again with pairs of 3-vertices on a single monad parameterized by D_2^4. There is no necessity that such 3-vertices generate entire tree level of the theory. The reason is the probable failure of the analogue of the BCFW recursion relations. One can restrict entire tree level to the recursion procedure and state that precisely this segment of the tree level is the 'physical' content of the theory. There remains, however, the question of Poincaré invariance of the segment. We are postponing the details of the analysis to the separate work not deciding here the final resolution.

Anyway, the pure perturbative and supersymmetric QG's are exactly solved by synthetic methods. These cases, as shown in this paper, teach us that EE are relating classical as well quantum fields, including quantized gravity. In this way one finds quantum nature of the classical $g_{\mu\nu}$ leading to finite results. The appearance of $g_{\mu\nu}$ in EE does not stay in the opposition to the quantized fields of matter since $g_{\mu\nu}$ is already quantized.

The existence of such finitely resolved cases of covariant quantized gravity encourages us to ask some general questions. They are related with the problem stated in the title of this section, i.e. how (if at all) it can be that a perturbative QG is considered a fundamental theory. Is this sensible at all? Forgetting for a while the problems with full matter QG on synthetic spacetime, let us think on the marriage of perturbative QG and synthetic methods applied to spacetime as giving the finite, well-defined results. Where can such approach be applied in physical reality? First, present cosmological models of inflation are based on the single scalar field ϕ theory of gravity with the action functional

$$\int d^4x \left[\frac{c^4}{16\pi G} R - \frac{1}{2}\phi_{,\mu}\phi_{,\nu}g^{\mu\nu} - V(\phi) \right]$$

where R is the Ricci scalar and $V(\phi)$ the potential which leads to the FRW geometry [18]

$$ds^2 = -c^2dt^2 + a^2(t)d\overline{x} \cdot d\overline{x}\,.$$

The quantization of the theory as the perturbative QG (with scalar matter field—the inflaton field) allows for calculating the tree, one loop and higher loops effects. The analysis and results are presented in [18]. It follows that the tree level contributes to the power spectra i.e. tensor-to-scalar ratio $r(k)$, the scalar spectral index $n_s(k)$ and the tensor spectral index $n_t(k)$. These parameters are determined nearly completely by the tree diagrams of the theory. Currently there are experimental constraints from PLANCK, WMAP and BICEP collaborations on the parameters which are known up to two significant figures. The near future data on CMB irregularities should allow for testing the tree predictions. These data could become first indications on QG effects ever, even though they would be related with the tree level of the scalar

GR. The one loop corrections to the spectra are calculable, in principle, in the scalar theory but their experimental verification would require 10 significant figures, which is eventually possible within next 20 years. Higher loop corrections are out of reach within predictable future (see e.g. [18]). The conclusion, which agrees also with synthetic predictions, is that the tree-level of the scalar theory suffices for covering the nearby QG results.

The tree level of QG reproduces also the classical content of the theory, hence $g_{\mu\nu}$ (in the limit) becomes again valid variable. Recalling this one finds stronger reasons indicating that *tree* canonical QG could be fundamental at any scale. In the previous sections we have shown that tree diagrams, hence classical $g_{\mu\nu}$, comprises the entire results of quantization of certain GR theories on synthetic spacetime. Such a result agrees with the approach to QG known as smooth QG [3] however, the classical $g_{\mu\nu}$ recovered after synthetic quantization undergoes some, rather subtle, changes. Smooth QG assumes that even on the Planck scales spacetime can be considered a smooth 4-manifold, hence $g_{\mu\nu}$ is the valid variable again. The smoothness structure is, however, induced by *exotic* smoothings of $\mathbb{R}^4$, i.e. nondiffeomorphic with the standard $\mathbb{R}^4$ and exoticness is essential for the approach. The existence of such exotic R^4's was established in mathematics in 1980s as the conclusion from breakthrough works of M. H. Freedman and S. K. Donaldson in differential topology and geometry. When exotic R^4 is used in the cosmological model [4], it leads to the realistic values of certain cosmological parameters, and can generate quantum algebras of observables [3]. Thus classical geometry of spacetime (exotic smoothness in dimension 4) can even precede the quantum regime of gravity (see also [14]).

The QG on synthetic spacetime approach developed in this paper shows that indeed the spacetime can be seen *almost* classical, i.e. with monads, even at deep Planck regime. This modification applied to the perturbative QG reproduces the tree level of the theory. On the other hand applying SDG methods to spacetime at small distances refers to the modified logic and set theory of a smooth topos $\mathscr{E}$. The question thus can be asked: Is this modified spacetime still any smooth manifold when seeing from SET perspective? The answer is YES, however, the smoothness structures of this 4-manifold can not be standard. This is precisely the subtle change witnessed in $g_{\mu\nu}$ mentioned above. In the case of $\mathbb{R}^4$ its synthetic deformation gives rise to exotic R^4 when projecting back to SET (e.g. [10, 11]). So we arrived at amazing coincidence between synthetic perturbative quantization and smooth quantization of gravity. What is more, this agreement and the entire construction can be successfully performed exclusively in dimension 4 where there exist exotic $\mathbb{R}^n$'s.

Summarizing, the approach via SDG to the perturbative QG sheds new light on the problem of quantizing gravity. The relation with smooth QG encourages us to consider synthetic perturbative QG a fundamental theory. It is fair to say that this is just the working hypotheses at the moment, nevertheless fascinating ones. Much theoretical work remains to be done along with confronting the results with the appearance of more and more refined experimental data.

Acknowledgements The author appreciates much all remarks made by Piotr Sułkowski regarding this paper since they improved its readability.

Appendix

A. Monads and Weil Algebras

The SET presentation of monads D_k, $k = 1, 2, \ldots$ is usually performed in terms of spectra of Weil algebras W, which are described as follows. Let W be a local ring which has $\mathbb{R}$-algebra structure. To be a local ring means that there exists unique maximal ideal m in W. *If W is a finite dimensional vector space over $\mathbb{R}$ and can be presented as $W = \mathbb{R} \oplus m$ then it is a Weil algebra.* We will see that m corresponds to D_k. The best way to understand it is via smooth ring presentation of Weil algebras. This is the content of the following theorem ([15, Theorem 3.17, p.37]).

Theorem 1 *Let W be a Weil algebra. This is equivalent to either of the following statements*

i. W is isomorphic to the $\mathbb{R}$-algebra $\mathbb{R}[X_1, X_2, \ldots, X_n]/I$ where I is the ideal of real polynomials containing all *powers of the degree k of X, i.e. for multiindices α, $\exists k \in \mathbb{N} \forall \alpha(|\alpha| = k, X^\alpha \in I)$.*
ii. W is isomorphic to a finite dimensional real vector space given by a ring $C_0^\infty(\mathbb{R}^n)/I$. Here $C_0^\infty(\mathbb{R}^n)$ is the ring of germs at 0, i.e. $C_0^\infty(\mathbb{R}^n) = C^\infty(\mathbb{R}^n)/m_g$.

By an abuse of notation (using internal to $\mathcal{E}$ object, D_k) one can write

$$\begin{aligned} C_0^\infty(\mathbb{R})/(X^2) &\simeq \mathbb{R} \oplus \varepsilon\mathbb{R} = \mathbb{R}[\varepsilon] = \mathbb{R} \oplus D_1 \\ C_0^\infty(\mathbb{R})/(X^3) &\simeq \mathbb{R} \oplus \varepsilon\mathbb{R} \oplus \varepsilon^2\mathbb{R} = \mathbb{R} \oplus D_2. \end{aligned} \tag{9.5}$$

Following [12] we can 'carve out' the subsets of polynomials in $\mathbb{R}[X]$ which, in terms of spectra of Weil algebras, lead to [12, p.43]

$$\begin{aligned} D &= Spec_{\mathbb{R}[X]}(C_0^\infty(\mathbb{R})/(X^2)) = \{b\varepsilon | \varepsilon^2 = 0, b \in \mathbb{R}\}, \\ D_2 &= Spec_{\mathbb{R}[X]}(C_0^\infty(\mathbb{R})/(X^3)) = \{b\varepsilon + c\varepsilon^2 | \varepsilon^3 = 0, b, c \in \mathbb{R}\}. \end{aligned} \tag{9.6}$$

Here $Spec_A(B)$ is the operation of carving out a subset of A by a ring $B = C^\infty(\mathbb{R})/(X^i)$, $i = 2, 3$.

The projection π, appearing in (9.3), can be extended over entire ring R as follows

$$\begin{aligned} \pi &: \mathcal{E} \to \mathrm{SET} \\ \pi &: R \mapsto C^\infty(\mathbb{R})/(x^3) = \mathbb{R} \oplus \varepsilon\mathbb{R} \oplus \varepsilon^2\mathbb{R}. \end{aligned} \tag{9.7}$$

Note that the natural part of the usual geometric morphism between SET and $\mathcal{E}$ sends R to $\mathbb{R}$. This modification of the geometric morphism in the context of gravity will be analyzed elsewhere.

B. The BCFW Recursion Relations and Tree Amplitudes of Gravity

Let us quote the fragment of [2] which exhibits the philosophy behind the on-shell 3-vertices diagrams with complex momenta as opposite to traditional Feynman diagrammatics:

Individual Feynman diagrams are not gauge invariant and thus don't have any physical meaning. By contrast, each on-shell diagram is physically meaningful and corresponds to some particular on-shell scattering process. Note that although on-shell diagrams almost always involve loops of internal particles, these internal particles often have momenta fixed by the constraints (or are otherwise free). On-shell forms are simply the products of on-shell 3-particle amplitudes; as such, they are always well-defined, finite objects, free from either infrared or ultraviolet divergences. This makes them ideal for exposing symmetries of a theory which are often obscured by such divergences.

Let $\mathbb{M}(p_1, \ldots, p_n, h_1, \ldots, h_n)$ be the S-matrix of a theory of gravity evaluated at n incoming momenta of gravitons with helicities h_i, $i = 1, \ldots, n$ (or other matter particles in a matter QG). $\mathbb{M}(p_1, \ldots, p_n, h_1, \ldots, h_n)$ is thus the amplitude for the process with these particles. The result is completely symmetric in all n external momenta. The BCFW procedure comprises complexified momenta $p_i(z) \in \mathbb{C}$, $i = 1, \ldots, n$ such that $\mathbb{M}(p_i, h_i)$ becomes a function of z, $\mathbb{M}(z)$, and such that all momenta are on-shell. This last requirement actually demands complex momenta. This goal is achieved by performing the special deformation (see e.g. [5]). The point of BCFW is the observation that all information about the analytical structure of $\mathbb{M}(z)$ is entirely coded in the simple poles. This follows from the general property of the analytical functions that any meromorphic function vanishing at infinity is determined by the simple poles. Crucial is the vanishing property $\lim_{z\to\infty} \mathbb{M}(z) = 0$ from which it follows

$$0 = \frac{1}{2\pi i} \int_C \frac{dz'}{z' - z} \mathbb{M}(z') = \mathbb{M}(z) + \{\text{residues}\}$$

when the contour C at infinity contains all the poles. Based on this the recursion relation for the arbitrary tree diagram amplitudes of on-shell momenta follows [1], and the BCFW relation reads

$$\mathbb{M}(z) = \sum_{L,h} \mathbb{M}_L(\{p_1(z_P), h_1\}, \{-P(z_P), h\}, L) \frac{1}{P^2(z)} \mathbb{M}_R(\{p_2(z_P), h_2\}, \{P(z_P), -h\}, R)\,. \tag{9.8}$$

Thus the BCFW general relation for arbitrary many n particles can be denoted schematically at the tree-level, as [6]

$$A_n = \sum_{r,h} A^h_{r+1} \frac{1}{P_r^2} A^{-h}_{n-r+1} \tag{9.9}$$

where $A^h_r(z) = \mathbb{M}(p_1(z), \ldots, p_n(z), h_1, \ldots, h_n)$ is the amplitude for r gravitons with helicities $h = (h_1, \ldots, h_n)$. All gravitons are on-shell with complex momenta and the minimal incoming data for the BCFW procedure are two 3-vertices glued by the propagator. This fact has been used by us in the text when indicating the similarity with the structure of gravity on synthetic spacetime. Importantly, the propagator $\frac{1}{P_r^2}$ for the graviton is on-shell as well and that is why we can freely switch between the gravitons and sum over r. The recursion formula for the amplitudes is not the procedure on Feynman diagrams leading to more and more complicated ones. It is rather the recursion formula on amplitudes of complex on-shell momenta. The power of this approach is this simple formula when compared to the extremely complicated world of Feynman diagrams [1].

References

1. N. Arkani-Hamed, F. Cachazo, J. Kaplan, What is the simplest quantum field theory? JHEP **1009**, 016 (2010)
2. N. Arkani-Hamed, J.L. Bourjaily, F. Cachazo, A.B. Goncharov, A. Postnikov, J. Trnka, Scattering amplitudes and the positive Grassmannian (2012). arXiv:1212.5605
3. T. Asselmeyer-Maluga, Smooth quantum gravity: exotic smoothness and quantum gravity, in *At the Frontier of Spacetime*, vol. 183, Fundamental Theories of Physics, ed. by T. Asselmeyer-Maluga (Springer, Switzerland, 2016)
4. T. Asselmeyer-Maluga, J. Król, How to obtain a cosmological constant from small exotic R^4. Phys. Dark Universe. **19**, 66–77 (2018), https://authors.elsevier.com/a/1WFBE7t6qpvmr8
5. P. Benincasa, C. Boucher-Veronneau, F. Cachazo, Taming tree amplitudes in general relativity. JHEP **11**, 057 (2007)
6. R. Britto, F. Cachazo, B. Feng, E. Witten, Direct proof of tree-level recursion relation in Yang-Mills theory. Phys. Rev. Lett. **94**, 181602 (2005)
7. M.J. Duff, Covariant quantization, in *Quantum Gravity. An Oxford Symposium*, 2nd edn., ed. by C.J. Isham, R. Penrose, D.W. Sciama (Clarendon Press, Oxford, 1978)
8. R.P. Feynman, F.B. Morinigo, W.G. Wagner, in *Feynman Lectures on Gravitation*, ed. by B. Hatfield (Addison Wesley, Reading, 1995)
9. R.P. Feynman, Quantum theory of gravitation. Acta Phys. Pol. **14**, 841 (1963)
10. M. Heller, J. Król, How logic interacts with geometry: Infinitesimal curvature of categorical spaces (2016). arXiv:1605.03099
11. M. Heller, J. Król, Infinitesimal structure of singularities. Universe **3**(1), 16 (2017)
12. A. Kock, *Synthetic Differential Geometry*, 2nd edn. (Cambridge University Press, Cambridge, 2006)
13. A. Kock, *Synthetic Geometry of Manifolds* (Cambridge University Press, Cambridge, 2009)
14. J. Król, T. Asselmeyer-Maluga, K. Bielas, P. Klimasara, From quantum to cosmological regime. The role of forcing and exotic 4-smoothness. Universe. **3**(2), 31 (2017)
15. I. Moerdijk, G.E. Reyes, *Models for Smooth Infinitesimal Analysis* (Springer, New York, 2010)
16. M.D. Scadron, *Advanced Quantum Theory*, 3rd edn. (Imperial College, World Scientific, New Jersey, 2007)
17. G. 't Hooft, Perturbative quantum gravity, in *Proceedings of the International School of Subnuclear Physics, Erice 2002, From Quarks and Gluons to Quantum Gravity*, ed. by A. Zichichi, Subnuclear Series, vol. 40 (World Scientific, 2002), pp. 249–269
18. R.P. Woodard, Perturbative quantum gravity comes of age. Int. J. Mod. Phys. D **23**(9), 1430020 (2014)

Chapter 10
Category Theory as a Foundation for the Concept Analysis of Complex Systems and Time Series

G. N. Nop, A. B. Romanowska and J. D. H. Smith

Abstract Wille's formal concept analysis, Hardegree's treatment of natural kinds, and Birkhoff's mathematical theory of polarities provide essentially equivalent tools for the analysis of a static system functioning at a single level. We now show how a number of categorical notions allow these tools to be extended to cover the analysis of complex systems involving multiple hierarchical levels indexed by a semilattice, including the case where a chain represents a time series governing the evolution of a single system. A semilattice is a poset category with finite products or coproducts. Our analysis then rests on functors from a semilattice to the category of complete lattice homomorphisms, or from a semilattice to a category of polarities and bonding relations.

10.1 Introduction

Concept analysis has emerged as a tool for the study of the relationships between selected items and the properties that they may possess. It identifies those groups of objects and related properties that embody significant features of the system under consideration. It is a static science that does not take account of the development of a changing system, or the progressive evolution of complexity. The fixed relationships that it identifies are exhibited by a concept lattice (Sect. 10.2).

The goal of the current work is to present the foundations for an extension of the ideas of concept analysis to changing environments, evolving complex systems, and time series analyses of successive stages of a given system. The extension relies

G. N. Nop · J. D. H. Smith
Iowa State University, Ames, IA 50011, USA
e-mail: gnnop@iastate.edu

J. D. H. Smith
e-mail: jdhsmith@iastate.edu

A. B. Romanowska (✉)
Warsaw University of Technology, Pl. Politechniki 1, 00-661 Warsaw, Poland
e-mail: a.romanowska@mini.pw.edu.pl

M. Kuś and B. Skowron (eds.), *Category Theory in Physics, Mathematics, and Philosophy*, Springer Proceedings in Physics 235,
https://doi.org/10.1007/978-3-030-30896-4_10

critically on the use of category theory in a number of ways. On the one side, the changing relationships between varying sets of items and properties are encoded within a category of polarities and bonding relations (Sect. 10.3). On the other side, we extend concept lattices to concept quasilattices (Sect. 10.5), whose structure is determined by contravariant functors from poset categories to the category of complete lattices (Sect. 10.4).

10.2 Polarities and Complete Lattices

Hardegree's treatment of natural kinds [4], Wille's formal concept analysis [2, 14], MacNeille's completions [7], and Birkhoff's theory of polarities [1, Sect. V.7] provide essentially equivalent tools for the analysis of a static system functioning at a single level. We summarize these tools in the present section.

10.2.1 Poset Categories

We will use the following terms:

- A *poset category* is a small category **C**, where for each pair x, y of objects of **C**, one has $|\mathbf{C}(x, y) \cup \mathbf{C}(y, x)| \in \{0, 1\}$[1];
- A *meet semilattice* is a poset category, in which each pair x, y of objects has a product $x \times y$;
- A *meet semilattice homomorphism* is a functor between meet semilattices that preserves products of pairs of objects;
- A *join semilattice* is a poset category, in which each pair x, y of objects has a coproduct $x + y$;
- A *join semilattice homomorphism* is a functor between join semilattices which preserves coproducts of pairs of objects;
- A *lattice* is a poset category, where each pair of objects has a product and a coproduct;
- A *lattice homomorphism* is a functor between lattices that preserves products and coproducts of pairs of objects—here **Lat** denotes the category of lattice homomorphisms between lattices;
- A *bounded lattice* is a lattice with an initial object $\bot$ and a terminal object $\top$;
- A *bounded lattice homomorphism* is a lattice homomorphism that preserves the initial and terminal objects;
- A *complete lattice* is a poset category which is *bicomplete*: each set S of objects has a product $\prod_{x\in S} x$ and a coproduct $\sum_{x\in S} x$;

[1]Recall that each pre-order or thin small category is categorically equivalent to a poset.

- A *complete lattice homomorphism* is a lattice homomorphism which preserves all products and coproducts. Then **CLat** denotes the category of complete lattice homomorphisms between complete lattices [6, Sect. I.4.3].

For objects x, y of a poset category P, the nonemptiness of the hom-set $P(x, y)$ will be denoted by $x \to y$ or $x \leq y$. In the latter notation, appropriate suffices may be attached, such as

$$x \leq_\times y \quad \text{corresponding to} \quad x = x \times y \tag{10.1}$$

in a meet semilattice, or

$$x \leq_+ y \quad \text{corresponding to} \quad x + y = y \tag{10.2}$$

in a join semilattice. Note that $\leq$ is a partial order relation on the object set (hence the term "poset"): reflexive, antisymmetric, and transitive.

10.2.2 Contexts

Suppose that Ω is a set of *items*, and that Π is a set of *properties*. Let α be a subset of $\Omega \times \Pi$, considered as a relation of *attribution* from Ω to Π. Thus if $(x, p) \in \alpha$ (or equivalently $x\ \alpha\ p$), property p is an attribute of item x. The triple (Ω, Π, α) is called a *polarity* [1, Sect. V.7] or *NK-structure* [4] or *context* [2, 14]. Equivalently, the disjoint union $\Omega \uplus \Pi$ may be construed as a poset (category), with nontrivial order relations $x \to p$ for $(x, p) \in \alpha$.

We list some classical examples of contexts:

Posets: For a poset $(H, \leq)$, take $(H, H, \leq)$.

Algebraic geometry: For a positive integer n, take $(\mathbb{C}^n, \mathbb{C}[X_1, \ldots, X_n], \alpha)$, where α is the *annihilation relation* with $(x_1, \ldots, x_n)\ \alpha\ f(X_1, \ldots, X_n)$ if and only if $f(x_1, \ldots, x_n) = 0$.

Galois theory: For a field extension $F \hookrightarrow E$, suppose that $[F, E]$ is the set of all intermediate fields. Let G be the *Galois group* of the extension, the group of automorphisms of E that fix F. For $K \in [F, E]$ and $g \in G$, write $K\ \alpha\ g$ if g fixes each element of K. Then $([F, E], G, \alpha)$ is a polarity.

10.2.3 Galois Connections and Galois Correspondences

For a set X of items, define

$$X^r = \{p \in \Pi \mid \forall\, x \in X\,,\ x\ \alpha\ p\} \tag{10.3}$$

as the set of properties common to all items in X. Dually, for a set P of properties, define

$$P^s = \{x \in \Omega \mid \forall\, p \in P\,,\ x\,\alpha\, p\} \qquad (10.4)$$

as the set of items attributed to all properties in P. The specifications (10.3) and (10.4) yield a pair

$$(2^\Omega, \subseteq) \underset{s}{\overset{r}{\rightleftarrows}} (2^\Pi, \supseteq) \qquad (10.5)$$

of order-preserving functions, an adjoint pair of functors between poset categories [13, III Sect. 3.3]. The pair (10.5) is known as a *Galois connection*. It restricts to a pair

$$(2^\Pi s, \subseteq) \underset{s}{\overset{r}{\rightleftarrows}} (2^\Omega r, \supseteq) \qquad (10.6)$$

of mutually inverse order-preserving functions, a *Galois correspondence*, between the respective images $2^\Omega r$ and $2^\Pi s$ of the functions r and s from (10.5). The terminology here relates to the context of Galois theory from Sect. 10.2.2.

10.2.4 Concepts

Consider a polarity or context (Ω, Π, α), with Galois correspondence (10.6). Elements of $2^\Omega r$ and $2^\Pi s$ are described as *closed*. Then a *concept* of the context is an ordered pair

$$(A|B) \qquad (10.7)$$

in $2^\Pi s \times 2^\Omega r$ (written using a vertical line as the separator), with $A^r = B$ and $B^s = A$. The closed set A is the *extent* of the concept (10.7), while the closed set B is the *intent* of the concept (10.7).

Example 1 In the algebraic geometry context of Sect. 10.2.2, the closed subsets of $\mathbb{C}^n$ are *algebraic sets*, while the closed subsets of $\mathbb{C}[X_1, \ldots, X_n]$ are *radical ideals* [5, Corollary I.1.4].

As a set, the *concept lattice* $\mathscr{L}(\Omega, \Pi, \alpha)$ is the set of all concepts of the context (Ω, Π, α). The *conceptualization* of an item $x \in \Omega$ is the concept

$$x^\varepsilon = (\{x\}^{rs}|\{x\}^r)\,, \qquad (10.8)$$

while the *conceptualization* of a property $p \in \Pi$ is the concept

$$p^\eta = (\{p\}^s|\{p\}^{sr})\,. \qquad (10.9)$$

Under the equivalent order relations

$$(A_1|B_1) \leq_\times (A_2|B_2) \quad \Leftrightarrow \quad A_1 \subseteq A_2 \tag{10.10}$$

and

$$(A_1|B_1) \leq_+ (A_2|B_2) \quad \Leftrightarrow \quad B_1 \supseteq B_2 \tag{10.11}$$

for $(A_1|B_1), (A_2|B_2) \in \mathscr{L}(\Omega, \Pi, \alpha)$, the concept lattice forms a complete lattice. Indeed, for a subset $\{(A_i|B_i) \mid i \in I\}$ of $\mathscr{L}(\Omega, \Pi, \alpha)$, the product is given by

$$\prod_{i \in I}(A_i, B_i) = \left(\bigcap_{i \in I} A_i \middle| \left(\bigcap_{i \in I} A_i\right)^r\right)$$

while

$$\sum_{i \in I}(A_i, B_i) = \left(\left(\bigcap_{i \in I} B_i\right)^s \middle| \bigcap_{i \in I} B_i\right)$$

gives the coproduct [1, V Theorem 19], [2, Satz/Theorem 3], [7]. We will write **CtLt** for the category of concept lattices and complete lattice homomorphisms between them. Note that there is an inclusion functor

$$\mathscr{A} : \mathbf{CtLt} \hookrightarrow \mathbf{CLat} \tag{10.12}$$

known as *abstraction*.

Example 2 In the poset context of Sect. 10.2.2, there is an order-preserving embedding

$$(H, \leq) \to \mathscr{L}(H, H, \leq); h \mapsto (h^{\geq}|h^{\leq}) \tag{10.13}$$

known as the *MacNeille completion* [7, Sect. 11]. The extent $h^{\geq} = \{k \in H \mid h \geq k\}$ and intent $h^{\leq} = \{k \in H \mid h \leq k\}$ correspond to the respective slice categories H/h and h/H. Note that $h^{\varepsilon} = h^{\eta} = (h^{\geq}|h^{\leq})$ using the conceptualization notation of (10.8) and (10.9). The MacNeille completion captures the construction of the ordered set $(\mathbb{R}, \leq)$ of real numbers from the ordered set $(\mathbb{Q}, \leq)$ of rational numbers.

10.2.5 Two-Element Posets

In order to obtain minimal nontrivial examples of concept lattices for subsequent use, we will examine the MacNeille completions (10.13) of the two two-element posets. The intents and extents of the concepts will be presented as concatenations of their elements.

Example 3 Consider the two-element lattice $\mathbf{2} = \{0 < 1\}$, which corresponds to the following poset context:

$$\begin{array}{c|cc} \leq & 0 & 1 \\ \hline 0 & \times & \times \\ 1 & & \times \end{array} . \tag{10.14}$$

Its MacNeille completion is the concept lattice $(0|01) \to (01|1)$, which is abstractly isomorphic to $\mathbf{2}$. Then

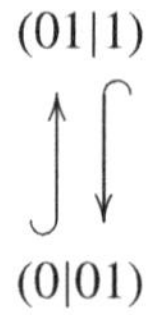

gives a set-theoretical representation of the respective order relations (10.10) and (10.11).

Example 4 Consider the two-element antichain $\{a, b\}$, which corresponds to the following poset context:

$$\begin{array}{c|cc} \leq & a & b \\ \hline a & \times & \\ b & & \times \end{array} . \tag{10.15}$$

Abstractly, its MacNeille completion is isomorphic to the lattice $\mathbf{2} \times \mathbf{2}$. Again,

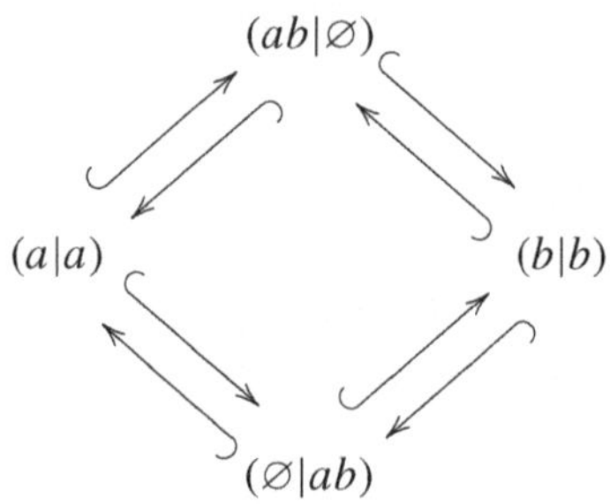

gives a set-theoretical representation of the respective order relations (10.10) and (10.11) on the concept lattice.

10.3 Categories of Concepts

In the preceding section, polarities or contexts were taken in isolation. We now examine certain relationships between contexts, obtaining various categories whose object class is the class of polarities or contexts.

10.3.1 The Category of Bonds

Consider a *domain context* $(\Omega_d, \Pi_d, \alpha_d)$ and a *codomain context* $(\Omega_c, \Pi_c, \alpha_c)$. Then a *bond* [2, Sect. 5.1] from the domain to the codomain is a relation $\beta_{dc} \subseteq \Omega_d \times \Pi_c$ such that:

- For each domain item x, the set

$$x^{\beta_{dc}} = \{q \in \Pi_c \mid x\ \beta_{dc}\ q\} \tag{10.16}$$

  is an intent in the codomain, and dually
- For each codomain property q, the set

$$^{\beta_{dc}}q = \{x \in \Omega_d \mid x\ \beta_{dc}\ q\} \tag{10.17}$$

  is an extent in the domain.

The notation of (10.16) and (10.17) is extended to subsets X of Ω_d and P of Π_c, so that

$$X^{\beta_{dc}} = \{p \in \Pi_c \mid \forall\, x \in X\,,\ x\ \beta_{dc}\ p\} = \bigcap_{x \in X} x^{\beta_{dc}}$$

and

$$^{\beta_{dc}}P = \{x \in \Omega_d \mid \forall\, p \in P\,,\ x\ \beta_{dc}\ p\} = \bigcap_{p \in P} {}^{\beta_{dc}}p\,.$$

For any domain context $(\Omega_d, \Pi_d, \alpha_d)$ and any codomain context $(\Omega_c, \Pi_c, \alpha_c)$, the relation $\Omega_d \times \Pi_c$ forms a bond from $(\Omega_d, \Pi_d, \alpha_d)$ to $(\Omega_c, \Pi_c, \alpha_c)$. For any context (Ω, Π, α), the polarity α is a bond from (Ω, Π, α) to (Ω, Π, α).

Now consider a further bond β_{dc} from $(\Omega_c, \Pi_c, \alpha_c)$ to a context $(\Omega_b, \Pi_b, \alpha_b)$. Then the relation

$$\beta_{dc}\beta_{cb} = \bigcup_{(A|B) \in \mathscr{L}(\Omega_c, \Pi_c, \alpha_c)} {}^{\beta_{dc}}B \times A^{\beta_{cb}} = \{(x, n) \in \Omega_d \times \Pi_b \mid (x^{\beta_{dc}})^s \subseteq {}^{\beta_{cb}}n\} \tag{10.18}$$

is a bond from $(\Omega_d, \Pi_d, \alpha_d)$ to $(\Omega_b, \Pi_b, \alpha_b)$ [2, Sect. 5.1]. The condition (10.18) on a pair (x, n) from $\Omega_d \times \Pi_b$ to be in $\beta_{dc}\beta_{cb}$ may be summarized by the following diagram:

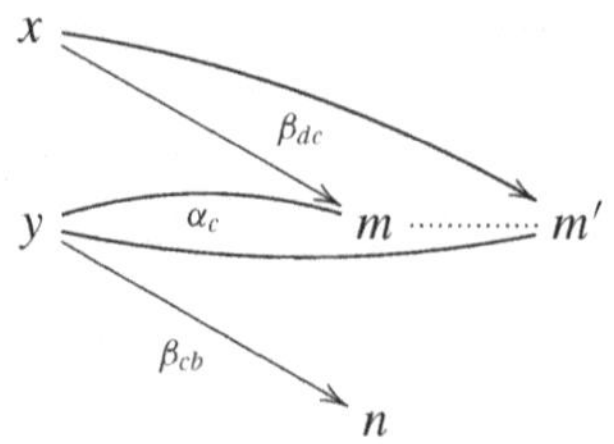

Here, the item y from Ω_c is α_c-related to all of the elements $m, \ldots, m'$ that are β_{dc}-related to x. Thus y represents a typical element of $(x^{\beta_{dc}})^s$. The diagram expresses the defining requirement (10.18) for such an element to lie in ${}^{\beta_{cb}}n$; namely, that it should be β_{cb}-related to n. This diagrammatic representation of the bond product may be contrasted with the representations used in [2, Sect. 5.1].

The *category* **Bond** *of bonds* has the class of contexts as its object class, and the class of bonds as its morphism class. The product of morphisms is given by (10.18). The identity morphism at a context (Ω, Π, α) is the bond α.

10.3.2 The Category of Bond Pairs

Suppose that $(\Omega_d, \Pi_d, \alpha_d)$ is a domain context and that $(\Omega_c, \Pi_c, \alpha_c)$ is a codomain context. Then a *bond pair* (β_{dc}, β_{cd}) from $(\Omega_d, \Pi_d, \alpha_d)$ to $(\Omega_c, \Pi_c, \alpha_c)$ consists of a bond β_{dc} from $(\Omega_d, \Pi_d, \alpha_d)$ to $(\Omega_c, \Pi_c, \alpha_c)$ and a bond β_{cd} from $(\Omega_c, \Pi_c, \alpha_c)$ to $(\Omega_d, \Pi_d, \alpha_d)$. Thus if (β_{dc}, β_{cd}) is a bond pair from $(\Omega_d, \Pi_d, \alpha_d)$ to $(\Omega_c, \Pi_c, \alpha_c)$, and that (β_{cb}, β_{bc}) is a bond pair from $(\Omega_c, \Pi_c, \alpha_c)$ to $(\Omega_b, \Pi_b, \alpha_b)$. Then

$$(\beta_{dc}, \beta_{cd})(\beta_{cb}, \beta_{bc}) := (\beta_{dc}\beta_{cb}, \beta_{bc}\beta_{cd}) \tag{10.19}$$

is a bond pair from $(\Omega_d, \Pi_d, \alpha_d)$ to $(\Omega_b, \Pi_b, \alpha_b)$.

The *category* **BdPr** *of bond pairs* comprises the class of contexts as its object class. The class of bond pairs is its morphism class. The product of morphisms is given by (10.19). Then the identity morphism at a context (Ω, Π, α) is the bond pair (α, α).

10.3.3 The Category of Bondings

Consider a given bond pair (β_{dc}, β_{cd}) from a domain context $(\Omega_d, \Pi_d, \alpha_d)$ to a codomain context $(\Omega_c, \Pi_c, \alpha_c)$. The bond pair is a *bonding* if the condition

$$\forall\,(A|B) \in \mathscr{L}(\Omega_d, \Pi_d, \alpha_d)\,,\ \left(A^{\beta_{dc}}\right)^r = {}^{\beta_{cd}}B \text{ and } \left({}^{\beta_{cd}}B\right)^s = A^{\beta_{dc}}$$

is satisfied.

For a domain context $(\Omega_d, \Pi_d, \alpha_d)$ and a codomain context $(\Omega_c, \Pi_c, \alpha_c)$, suppose that

$$\varphi\colon \mathscr{L}(\Omega_d, \Pi_d, \alpha_d) \to \mathscr{L}(\Omega_c, \Pi_c, \alpha_c)$$

is a complete lattice homomorphism. Define

$$\beta_{dc} = \{(x, q) \in \Omega_d \times \Pi_c \mid x^{\varepsilon}\varphi \leq q^{\eta}\} \tag{10.20}$$

and

$$\beta_{cd} = \{(y, p) \in \Omega_c \times \Pi_d \mid y^{\varepsilon} \leq p^{\eta}\varphi\} \tag{10.21}$$

using the conceptualization notation of (10.8) and (10.9). Then (β_{dc}, β_{cd}) is a bonding from $(\Omega_c, \Pi_c, \alpha_c)$ to $(\Omega_d, \Pi_d, \alpha_d)$ [2, Sect. 7.2]. We thus obtain a functor **CtLt** $\to$ **BdPr** (where the functoriality follows along the lines of [2, Hilfs./Proposition 113].) The image of this functor yields a *category of bondings* **Bdg** as a subcategory of **BdPr**.

Conversely, for a domain context $(\Omega_d, \Pi_d, \alpha_d)$ and a codomain context $(\Omega_c, \Pi_c, \alpha_c)$, suppose that there is a bonding (β_{dc}, β_{cd}) from the domain to the codomain. Then

$$\mathscr{L}(\Omega_d, \Pi_d, \alpha_d) \to \mathscr{L}(\Omega_c, \Pi_c, \alpha_c)\colon (A|B) \mapsto \left({}^{\beta_{cd}}B \middle| A^{\beta_{dc}}\right)$$

defines a complete lattice homomorphism [2, Sect. 7.2]. We obtain an isomorphism

$$\mathscr{L}\colon \mathbf{Bdg} \to \mathbf{CtLt} \tag{10.22}$$

between the category **Bdg** of bondings and the category **CtLt** of complete lattice homomorphisms between concept lattices.

10.3.4 A Complete Lattice Homomorphism

The diagonal embedding $\Delta\colon \mathbf{2} \to \mathbf{2} \times \mathbf{2}$ of the two-element lattice into its direct square is a complete lattice homomorphism. Interpreting the domain and codomain of Δ as the respective concept lattices presented in Sect. 10.2.5, we have

$$\Delta\colon (01|1) \mapsto (ab|\varnothing)\,,\ (0|10) \mapsto (\varnothing|ab)$$

as the action of the complete lattice homomorphism. The corresponding bonding (β_{dc}, β_{cd}) is given by

$$\beta_{dc} = \{(0, a), (0, b)\}$$

recording that the image of 0^{ε} is below a^{η} and b^{η}, according to (10.20), and

$$\beta_{cd} = \{(a, 1), (b, 1)\}$$

recording that a^{ε} and b^{ε} are below the image of 1^{η}, according to (10.21). The bonding may be displayed as

	0	1	a	b
0	×	×	×	×
1		×		
a		×	×	
b		×		×

(10.23)

in conjunction with the contexts from Examples 3 and 4.

10.4 Quasilattices

A *bisemilattice* $(Q, \times, +)$ consists of a set Q that forms both a meet semilattice $(Q, \leq_{\times})$ with products $x \times y$ and a join semilattice $(Q, \leq_{+})$ with coproducts $x + y$, for x, y in Q. A *bisemilattice homomorphism* from a bisemilattice $(Q, \times, +)$ to a bisemilattice $(Q', \times, +)$ is a function $f \colon Q \to Q'$ such that $f \colon (Q, \leq_{\times}) \to (Q', \leq_{\times})$ is a meet semilattice homomorphism and $f \colon (Q, \leq_{+}) \to (Q', \leq_{+})$ is a join semilattice homomorphism.

A bisemilattice $(Q, \times, +)$ is a *quasilattice* if

$$[(x + y) \times z] + [y \times z] = (x + y) \times z$$

and

$$[(x \times y) + z] \times [y + z] = (x \times y) + z$$

for elements x, y, z of Q. Under an additional distributivity assumption, quasilattices were introduced by Płonka [9]. They were subsequently considered in full generality by Padmanabhan [8]. Their duality was studied initially in the distributive case [3], and more recently in the general case [12].

10.4.1 Lattices and Semilattices as Quasilattices

Since a quasilattice consists of two poset categories sharing a common object set, we may identify two important classes of quasilattices:

- A quasilattice $(Q, \times, +)$ is a **lattice** if its two poset category structures coincide;
- A quasilattice $(Q, \times, +)$ is a **semilattice** if its two poset category structures are mutually dual.

Thus a quasilattice $(Q, \times, +)$ is a lattice if either of the two *absorption* conditions

$$x + (x \times y) = x \quad \text{or} \quad x \times (x + y) = x$$

(expressing the equivalence of $\leq_\times$ and $\leq_+$) holds for x, y in Q. On the other hand, it is a semilattice if $x \times y = x + y$ (expressing the equivalence of $\leq_\times$ and $\geq_+$) holds for x, y in Q.

10.4.2 The Structure of Quasilattices

Consider the categories **SLat** and **QLat** of bisemilattice homomorphisms, taken with respective object classes consisting of semilattices and quasilattices. Then the nameless inclusion functor $\mathbf{SLat} \hookrightarrow \mathbf{QLat}$ has a left adjoint $P\colon \mathbf{QLat} \to \mathbf{SLat}$, the *reflection* or *replica* functor. The image Q^P of a quasilattice Q is its *semilattice replica*. Write $\pi : Q \to Q^P$ for the unit of the adjunction, interpreted as an analogue of a fibre bundle over a base space Q^P, with Q as the total space of the bundle. (This analogy is made more explicit in [12, Sect. 5.1.3].) Then for each point h in $H = Q^P$, the *Płonka fibre* $h^R = \pi^{-1}\{h\}$, as a subquasilattice of Q, is actually a lattice. If $h \leq_\times k$ in Q^P, there is a uniquely defined lattice homomorphism

$$(h \leq_\times k)^R \colon \pi^{-1}\{k\} \to \pi^{-1}\{h\};\, x \mapsto x + (x \times y)$$

that is independent of the choice of an arbitrary element y of $\pi^{-1}\{h\}$. We call these lattice homomorphisms the *Płonka homomorphisms* of the quasilattice $(Q, \times, +)$. Thus the quasilattice Q serves to specify a contravariant functor $R\colon (H, \leq_\times) \to \mathbf{Lat}$ from the poset category $(H, \leq_\times)$, the semilattice replica of Q, to the category **Lat** of lattices.

Conversely, suppose that there is a contravariant functor $R\colon (H, \leq_\times) \to \mathbf{Lat}$ from a meet semilattice $(H, \leq_\times)$ to the category **Lat** of lattices. A quasilattice structure is defined on the disjoint union $\biguplus_{h \in H} h^R$ by

$$x \times y = x(h \times k \leq_\times h)^R \times y(h \times k \leq_\times k)^R$$

and

$$x + y = x(h \times k \leq_\times h)^R + y(h \times k \leq_\times k)^R$$

for $h, k \in H$, along with $x \in h^R$ and $y \in k^R$. These two constructions provide an equivalence between quasilattices Q and corresponding *representation functors* $R\colon Q^P \to \mathbf{Lat}$ [8, 10], [11, Theorem 4.3.2], [12, Sect. 6.1]. In the literature, contravariant functors from $(H. \leq_\times)$ are sometimes taken in the form of covariant functors from $(H. \leq_+)$.

10.4.3 The Diagonal Embedding of the Two-Element Lattice

In Sect. 10.3.4, we examined the diagonal embedding $\Delta\colon \mathbf{2} \to \mathbf{2} \times \mathbf{2}$ of the two-element lattice $\mathbf{2} = \{0 \leq 1\}$ into its direct square. The lattice homomorphism may be displayed as

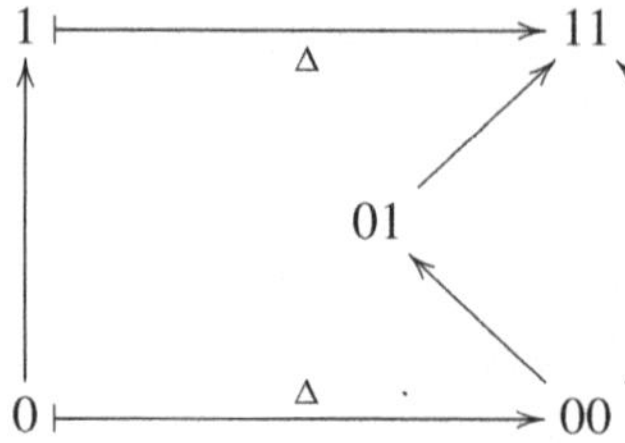

(10.24)

using a convention where the elements of the direct square are written simply as binary strings of length two. Note that the lattice structures are presented by their *Hasse diagrams*, the directed graphs recording the covering relations in the lattices. In other words, the full order relations on the lattices are the reflexive, transitive closures of the covering relations displayed in the Hasse diagram.

Now consider a two-element meet semilattice

$$H = \{\texttt{early} \to \texttt{late}\} \tag{10.25}$$

which readers are invited to interpret as an encoding of a time interval. Consider a contravariant functor $R\colon H \to \mathbf{Lat}$. Consider the representation $R\colon H \to \mathbf{Lat}$ with $\texttt{early}^R = \mathbf{2} \times \mathbf{2}$, $\texttt{late}^R = \mathbf{2}$, and $(\texttt{early} \to \texttt{late})^R = \Delta$. The construction from Sect. 10.4.2 then yields a quasilattice $(Q, \times, +)$ with

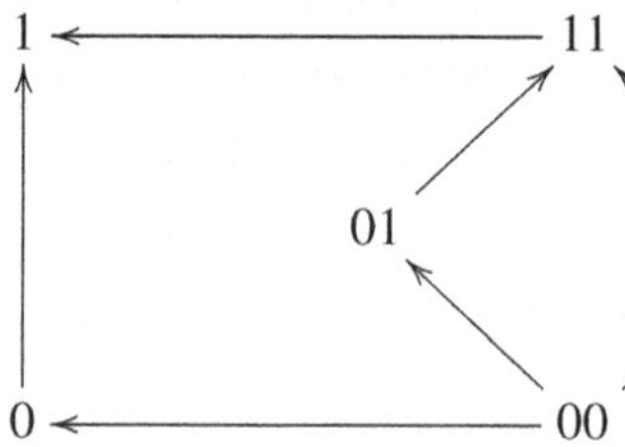

(10.26)

as the Hasse diagram of $(Q, \leq_\times)$ and

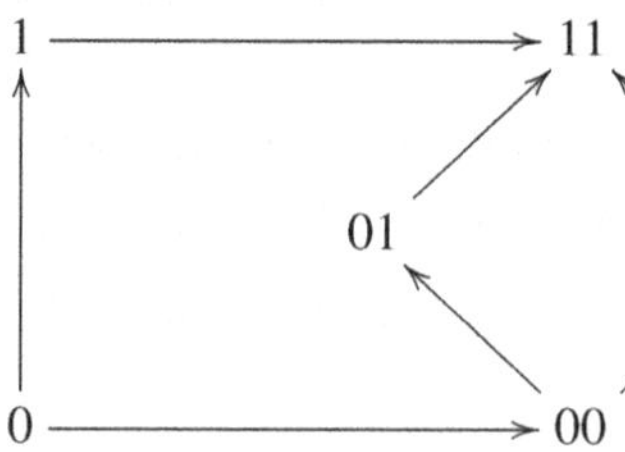

(10.27)

as the Hasse diagram of $(Q, \leq_+)$. These diagrams illustrate features common to all quasilattices. For example, two elements lie in the same lattice fibre if and only if they have the same order relationship in each diagram. On the other hand, an order relationship under $\leq_+$ records the action of a lattice homomorphism if the dual relationship holds under $\leq_\times$.

10.5 Concept Analysis of a Complex System

We are now in a position to extend the ideas of concept analysis to complex systems with a number of distinct levels that are indexed by a semilattice. The semilattice may be representing a hierarchical arrangement of levels in a complex system. As an alternative, in the case in which the underlying semilattice is a chain, it may be interpreted as presenting a series of time points. A quasilattice with a semilattice replica of this type may be describing the history of a context, with its concomitant concept lattice, that is evolving over the time series. These quasilattices may be considered as dynamic versions of concept lattices.

In general, regardless of whether the underlying semilattice is a chain or not, it turns out to be convenient to use a temporal terminology ("earlier," "later," etc.), as in the protoplasmic example (10.25). In fact, as complex structures evolve, the higher levels in their hierarchy tend to arise later in the evolution, emerging on the basis of lower, earlier levels.

10.5.1 Complex Polarities

We first extend the ideas of Sect. 10.2.2 to complex systems with a hierarchy of distinct levels, indexed by a semilattice. Define a *complex polarity* or *quasicontext* to be a contravariant functor $P \colon H \to \mathbf{Bdg}$ from a semilattice to the category of bondings.

Example 5 The polarities and bonding displayed in (10.23) may be interpreted as a complex polarity $P \colon H \to \mathbf{Bdg}$ from the time interval semilattice H of (10.25):

- The image of the object `early` is the poset context (10.15) from Example 4;
- The image of the object `late` is the poset context (10.14) from Example 3;
- The image of the morphism `early` $\to$ `late` is the bonding from Sect. 10.3.4.

Note that polarities or contexts in the traditional sense are complex polarities or quasicontexts for which the indexing semilattice H is a singleton.

10.5.2 Concept Quasilattices

Now we may extend the ideas of Sect. 10.2.4 to complex systems with a hierarchy of distinct levels, indexed by a semilattice. A preliminary mathematical definition is required: A quasilattice is said to be *locally complete* if its Płonka fibres are actually complete lattices, and its Płonka homomorphisms are complete lattice homomorphisms. Thus under the equivalence presented in Sect. 10.4.2, locally complete quasilattices correspond to contravariant representation functors $R\colon (H, \leq_\times) \to \mathbf{CLat}$ from a meet semilattice $(H, \leq_\times)$ to the category **CLat** of complete lattices. In particular, both complete lattices and arbitrary semilattices form locally complete quasilattices.

Recall the isomorphism functor $\mathscr{L}$ of (10.22) and the abstraction functor $\mathscr{A}$ of (10.12). Now suppose that $P\colon H \to \mathbf{Bdg}$ is a complex polarity or quasicontext, as introduced in Sect. 10.5.1. Then the *concept quasilattice* of P is the locally complete quasilattice that is given by the representation

$$H \xrightarrow{P} \mathbf{Bdg} \xrightarrow{\mathscr{L}} \mathbf{CtLt} \xrightarrow{\mathscr{A}} \mathbf{CLat} \tag{10.28}$$

of the semilattice H in the category **CLat** of complete lattices.

Remark 1 If the semilattice H is a singleton $\{h\}$, the image of h under the functor P is a context. Then the image of h under (10.28) is the concept lattice of that context. This concept lattice is the quasilattice represented by the functor (10.28).

As a disjoint union, the concept quasilattice of P consists of all the concepts from its individual lattice fibres. For each indexing element h in the semilattice H, the object hP of **Bdg** is a context, whose concepts then arise first in the concept lattice $HP\mathscr{L}$. The order relationships between the various concepts of the concept quasilattice may be derived formally from the quasilattice structure, by means of (10.1) or (10.2). However, it is more illuminating to combine the set representations of the individual concept lattices of the contexts hP as h ranges over H. This process is illustrated and discussed below.

10.5.3 Set Representations: The Layers of History

Consider the complex polarity of Example 5. Abstractly, its concept quasilattice is given by the representation $R\colon \{\texttt{early} \to \texttt{late}\} \to \mathbf{Lat}$ of Sect. 10.4.3, which may be read as a contravariant functor $R\colon \{\texttt{early} \to \texttt{late}\} \to \mathbf{CLat}$, in view of the fact that $\Delta\colon \mathbf{2} \to \mathbf{2} \times \mathbf{2}$ is actually a complete lattice homomorphism. The respective meet semilattice and join semilattice structures of the abstract quasilattice are given by the Hasse diagrams (10.26) and (10.27). These Hasse diagrams then appear within the set representation of the concept quasilattice that is displayed in Fig. 10.1. Here, the separate meet-semilattice and join-semilattice order relationships between concepts correspond to containments of extents and intents, exactly as in the classical concept lattice specifications (10.10) and (10.11).

Note that the containments are opposed within the lattice fibres, as is usual for concept lattices. On the other hand, the containments actually coincide along the

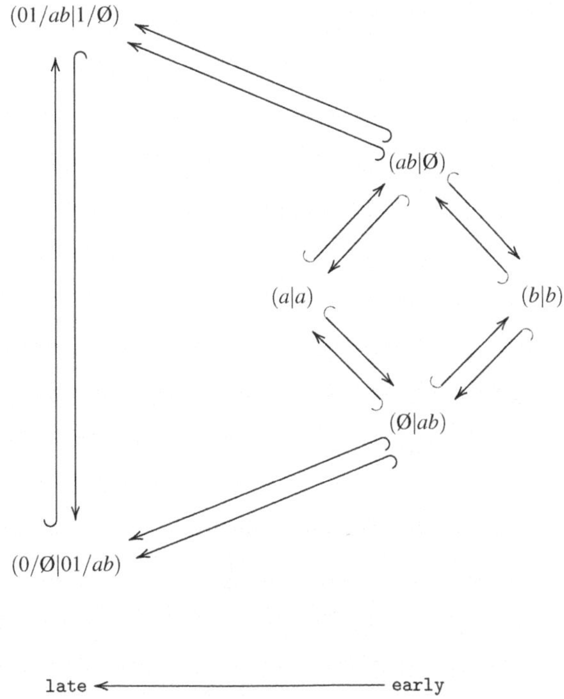

Fig. 10.1 The set representation of a concept quasilattice as a record of history

edges in the Hasse diagrams giving the actions of the lattice homomorphism that are displayed by barred arrows in (10.24). In order to implement these containments, the extents and intents of the domain concepts are augmented by the respective extents and intents of their images in the codomain concept lattice. In the figure, the augmented domain concepts are separated by a slash, suggesting that the later concepts appear on top of the corresponding earlier concepts in so-called *layers of history*.

In the time-series interpretation of quasicontexts, these layers are understood to be analogues of geological strata, or the successive layers of an archaeological site. In the complex system interpretation of quasicontexts, the upper layers represent advanced or high-level phenomena that have emerged from more primitive, low-level phenomena.

Looking at Fig. 10.1, it may be seen that the early full and empty intents and extents ab and $\varnothing$ are preserved over the time period, appearing as lower layers of history in the late period. On the other hand, the singleton extents and intents a, b are lost at that later time, corresponding to the fact that they are not part of the image of the complete lattice homomorphism in the structural representation of the concept quasilattice.

10.6 Conclusion

We have extended concept analysis to complex systems with multiple hierarchical levels, and to time series of evolving contexts. In particular, concept lattices have been extended to concept quasilattices. Within our theory, set representations of concept quasilattices provide a model for the accretion of structure that is preserved over history, and the disappearance of structure that fails to survive.

References

1. G. Birkhoff, *Lattice Theory*, 3rd edn. (American Mathematical Society, Providence, 1967)
2. B. Ganter, R. Wille, *Formale Begriffsanalyse* (Springer, Berlin, 1996) Translated by C. Franzke as Formal Concept Analysis (Springer, Berlin, 1999)
3. G. Gierz, A. Romanowska, Duality for distributive bisemilattices. J. Aust. Math. Soc. **51**, 247–275 (1991)
4. G.M. Hardegree, An approach to the logic of natural kinds. Pac. Phil. Q. **63**, 122–132 (1982)
5. R. Hartshorne, *Algebraic Geometry* (Springer, Berlin, 1977)
6. P.T. Johnstone, *Stone Spaces* (Cambridge University Press, Cambridge, 1982)
7. H.M. MacNeille, Partially ordered sets. Trans. Am. Math. Soc. **42**, 416–460 (1937)
8. R. Padmanabhan, Regular identities in lattices. Trans. Am. Math. Soc. **158**, 179–188 (1971)
9. J. Płonka, On distributive quasilattices. Fundam. Math. **60**, 191–200 (1967)
10. J. Płonka, On a method of construction of abstract algebras. Fundam. Math. **61**, 183–189 (1967)
11. A.B. Romanowska, J.D.H. Smith, *Modes* (World Scientific, River Edge, 2002)
12. A.B. Romanowska, J.D.H. Smith, Duality for quasilattices and Galois connections. Fund. Inf. **156**, 331–359 (2017)
13. J.D.H. Smith, A.B. Romanowska, *Post-Modern Algebra* (Wiley, New York, 1999)
14. R. Wille, Restructuring lattice theory: an approach based on hierarchies of concepts. In Ordered sets (Banff, AB, 1981), NATO Adv. Study Inst. Ser. C: Math. Phys. Sci. vol. 83 (Reidel, Dordrecht, 1982) pp. 445–470

The manufacturer's authorised representative in the EU is Springer Nature Customer Service Centre GmbH, Europaplatz 3, 69115 Heidelberg, Germany. If you have any concerns regarding our products, please contact ProductSafety@springernature.com

Printed and bound by CPI Group (UK) Ltd, Croydon, CR0 4YY

22/09/2026

02221061-0017